THE BODY IN TIME

The Wiley Science Editions

The Search for Extraterrestrial Intelligence, by Thomas R. McDonough

Seven Ideas that Shook the Universe, by Bryon D. Anderson and Nathan Spielberg

The Naturalist's Year, by Scott Camazine

The Urban Naturalist, by Steven D. Garber

Space: The Next Twenty-Five Years, by Thomas R. McDonough

The Body in Time, by Kenneth Jon Rose

Clouds in a Glass of Beer, by Craig Bohren

The Complete Book of Holograms, by Joseph Kasper and Steven Feller

The Scientific Companion, by Cesare Emiliani

Starsailing, by Louis Friedman

THE BODY IN TIME

Kenneth Jon Rose

Wiley Science Editions

JOHN WILEY & SONS, INC.

New York • Chichester • Brisbane • Toronto • Singapore

Publisher: Stephen Kippur
Editor: David Sobel
Managing Editor: Ruth Greif
Production: Publications Development Company of Texas
Illustrator: Jana Brenning

Library of Congress Cataloging-in-Publication Data

Rose, Kenneth Jon.
The body in time.

(Wiley science editions)
1. Chronobiology. I. Title. II. Series.
QP84.6.R67 1988 612'.022 87-21666
ISBN 0-471-85762-9

Printed in the United States of America

88 89 10 9 8 7 6 5 4 3 2 1

For Mom,
doctor, lawyer, chef,
and hired gun

There are so many facinating facts about time and the body that it was impossible to include everything in one book. Some of the unique and useful details are provided as asides in this format throughout the book.

ACKNOWLEDGMENTS

When I began this book, I naively thought that the trail set before me was going to be a smooth and level one. After all, how difficult could it be to find information on the duration of a sneeze, or why puberty begins when it does?

But, as I dove into the literature, it became readily apparent that the time parameters for a number of important events in the human body would be hard to find. Most textbooks don't include such information. And the current scientific literature usually cites earlier studies and assumes that the reader is already intimately familiar with the background. (I wasn't.) Gradually, though, whether it was from fingering through obscure journals or digging between the dusty covers of forgotten books, I found what I was looking for.

The task was enormous indeed. Luckily, I had the assistance of the librarians at New York University's Bobst Library, at the Library at New York University's Medical Center, and at the New York City Public Library. When I was about to throw up my arms and escape for home, they rescued me. These literary detectives would smile and hand me just the right book for my work. I am grateful to all of them.

I would also like to thank New York University Professors Walter Scott, William Crotty, Irving Brick, Gary Aston-Jones, and Efrain Azmitia for their help and encouragement throughout the years, and for investing their time to advance the career of a lowly graduate student.

I am indebted to all of my colleagues at New York University, as well. Specifically, I would like to thank Judith Gibber for her suggestions on primate behavior and pandas, Robert

Preti for his wise council, Maryann Huie for her input on blood and the immune system and for her fresh coffee and conversation, Vince Pieribone and Matt Ennis for their advice on the brain, and Neal Azrolan for his art. I offer my most special thanks to Martha Davila and the ladies on the ninth floor for their moral support and their warm smiles. Annabell Segarra, Ruth Frischer, and Jean King, of course, have my deepest gratitude. They made life in the laboratory a pleasure (and even laughed at my jokes). I shall value their friendship always.

My fondest regards will forever go to Ronit M. for her unflagging support, and for her love.

The manuscript itself could not have been written without the exacting input and extremely helpful suggestions of my editor, David Sobel, whose fingers were always either wrapped around a pencil, or around the phone, dialing me. He patiently read every draft and called my attention to many lapses in logic and continuity. I am indebted to him.

I would also like to acknowledge Jana Brenning for her wonderful line drawings that introduce each chapter, Ruth Greif for shepherding the book through publication, Nancy Marcus Land for doing such a great job in the production of the book, and Corinne McCormick at Wiley for her outstanding efforts in gathering the illustrations for the book.

I would also like to thank my agent, Barbara Bova, and her husband Ben for, well, everything. They know what I mean.

While I'm at it, I would like to send my appreciation and deepest love to Claudine, Skip, Cathy, and Stephanie Marquet, my second family. They were there at the beginning when I was but a poor student in a beat up red Toyota looking for a place to stay.

Finally, my deepest debt of gratitude goes to Dr. Fleur Strand. Her support, guidance, and especially her patience during this long ordeal, were invaluable. I dare say that I could not have completed this book without her help. More significantly, perhaps, it was from watching her that I acquired the skills of a scientist and the ingenuity of a teacher. And for that, grateful is too weak a word.

K. J. R.
New York City

PERMISSIONS

Page 5: T. W. Torrey and A. Feduccia, *Morphogenesis of the Vertebrates,* 4th ed., copyright © 1979 by John Wiley & Sons, Inc. Reprinted by permission of John Wiley & Sons, Inc.

Page 23: Kimble, Garmezy, Zigler, *Principles of Psychology,* 6th ed., copyright © 1984 by John Wiley & Sons, Inc. Reprinted by permission of John Wiley & Sons, Inc.

Page 31: T.W. Torrey and A. Feduccia, *Morphogenesis of the Vertebrates,* 4th ed., copyright © 1979 by John Wiley & Sons, Inc. Reprinted by permission of John Wiley & Sons, Inc.

Page 58: Burke, *Human Biology in Health and Disease,* copyright © 1975 by John Wiley & Sons, Inc. Reprinted by permission of John Wiley & Sons, Inc.

Page 60: F.L. Strand. *Physiology: A Regulatory Systems Approach,* 1st ed., copyright © 1978 by F.L. Strand. Reprinted by permission of MacMillan Publishing Co., Inc.

Page 71: F.L. Strand. *Physiology: A Regulatory Systems Approach,* 1st ed., copyright © 1978 by F.L. Strand. Reprinted by permission of MacMillan Publishing Co., Inc.

Page 77: Drawing by Neal Azrolan.

Page 78: Thorpe, *Cell Biology,* copyright © 1984 by John Wiley & Sons, Inc. Reprinted by permission of John Wiley & Sons, Inc.

Page 93: Wellcome Institute Library, London.

Page 101: Wellcome Institute Library, London.

Page 104, 105: Wellcome Institute Library, London.

Page 118: F.L. Strand. *Physiology: A Regulatory Systems Approach,* 1st ed., copyright © 1978 by F.L. Strand. Reprinted by permission of MacMillan Publishing Co., Inc.

Page 124: Wellcome Institute Library, London.

Page 126: Harlow Primate Lab, University of Wisconsin.

Page 127: Grobstein, *External Human Fertilization,* copyright © 1979 by Scientific American, Inc. All rights reserved.

Page 130: Patten, *Human Embryology,* 3rd ed., copyright © 1968 by McGraw-Hill Book Company.

Page 134: Wellcome Institute Library, London.

Page 149: Kimble, Garmezy, Zigler, *Principles of Psychology,* 6th ed., copyright © 1984 by John Wiley & Sons, Inc. Reprinted by permission of John Wiley & Sons, Inc.

Page 153: By permission of the Burndy Library, Norwalk, Connecticut.

Page 158: Worcester Art Museum, Worcester, Massachusetts.

Page 191: Kallmann & Jarvik, in Birren (ed), *Handbook of Aging and the Individual,* figure 8, copyright © 1959. Reprinted by permission of The University of Chicago Press.

Page 197: Courtesy of James D. Watson.

Page 199: Brady, Humiston, *General Chemistry,* 4th ed., copyright © 1986 by John Wiley & Sons, Inc. Reprinted by permission of John Wiley & Sons, Inc.

Page 201: F.L. Strand. *Physiology: A Regulatory Systems Approach,* 1st ed., copyright © 1978 by F.L. Strand. Reprinted by permission of MacMillan Publishing Co., Inc.

Page 205: F.L. Strand. *Physiology: A Regulatory Systems Approach,* 1st ed., copyright © 1978 by F.L. Strand. Reprinted by permission of MacMillan Publishing Co., Inc.

Page 208: Kimble, Garmezy, Zigler, *Principles of Psychology,* 6th ed., copyright © 1984 by John Wiley & Sons, Inc. Reprinted by permission of John Wiley & Sons, Inc.

Page 210: Kimble, Garmezy, Zigler, *Principles of Psychology,* 6th ed., copyright © 1984 by John Wiley & Sons, Inc. Reprinted by permission of John Wiley & Sons, Inc.

CONTENTS

"To everything there is a season and
a time to every purpose under the heaven . . ."

(Ecclesiastes, ch. 3, 1–8)

ONE

TIMING IS EVERYTHING

Once upon a time, there lived humans who were cursed with incredibly bad timing. They were very easy to recognize.

In a fist fight, they were the ones who swung and missed, only to be flattened by a powerful right upper cut to the jaw from their opponent. After a battle with sword and spear, they were the ones who were called "pegleg," or "lefty," or "one-eye." In a duel with pistols, they were the ones who died.

Like all groups, they had their famous members. Goliath was one (he didn't duck in time). Alexander Hamilton, one of this country's founding fathers, was another.

Early in the morning, on the eleventh of July, 1804, on the heights of Weehawken, New Jersey, on the same spot where Hamilton's eldest son, Philip, had died in a duel three years before, Hamilton and his antagonist, Aaron Burr met to end an argument. Three months before, the former Secretary of the Treasury had told all who would listen that he held a "despicable opinion" of Burr. Now, the Vice President demanded satisfaction for the inflammatory remarks about his character. So, as a man of honor, Hamilton felt himself compelled to accept Burr's challenge. It was not a good idea.

Burr was well-trained in the use of the dueling pistol. But, Hamilton, who held an aversion to the practice of dueling, was very rusty. So it was, on that fateful morning in July, that Burr and Hamilton faced each other, raised their pistols, aimed, and fired. Unfortunately for the country, Burr was faster and more accurate than Hamilton, and won the duel. His bullet found its mark and Hamilton fell. "This is a mortal wound, Doctor," the young statesman said to the physician at the scene. He died the following day.

A little more than one hundred years later, duels were being fought in the skies over Europe. Before 1914, before there were machine guns mounted on biplanes, pilots fired at each other with rifles and pistols. But taking aim from their excessively vibrating crafts proved to be impossible. They nearly always missed their targets.

Then someone came up with the bright idea that enemy planes could be destroyed in the air by dropping grenades onto them from planes above. But a pilot's timing had to be awfully good for it to work. If a pilot let go of the grenade too early, the shell might explode too far from the enemy plane to do any good. If a pilot held onto the ticking explosive just a smidgen too long, it might detonate just below his own plane, or worse, in his hand. Which is why it was always easy to tell who was good at this technique and who wasn't. The successful pilots came back with smiles on their faces; the others didn't come back at all.

Nowadays, the "poor timing" types seem to be everywhere. They're the ones who swing late and hit the tennis ball over the fence, or tell a joke then flub the punchline. They're the ones who slam on the brakes too late and hit the fender of the car ahead of them. They're the ones who do poorly on multiple-choice exams because they did not have enough time to complete them.

What is wrong with these people?

The Heart of the Issue

It depends on whom you ask. To the psychologist, timing is part of a mental awareness of the passage of time. The successful grenade thrower, for example, subconsciously "knows" when to toss the explosive at the right time. Psychologists would say that he does this by somehow mentally ticking off the duration in his head.

But all of us have this ability to measure the passage of time as well. Because of this gift, we can wake up moments before the morning alarm goes off, or we can get to the dryer seconds before the laundry is finished. I'm aware of this aptitude each time I hear a leaky faucet. It's the reason why waiting for the next plunk of water after listening to *drip* . . . *drip* . . . *drip* . . . . *drip* . . . , can be so maddening. Anticipation is the consequence of mental timing.

If psychologists see timing this way, biologists do not. To the biologist, timing is activity, movement, development, speed, and growth. To the biologist, every organ in the body, every tissue and cell, *every molecule,* performs its function within a measurable duration. To the biologist, each of the activities of the body has a noticeable and almost predictable start and finish.

Scientists, for example, expect a cell that lines the inside of the intestine to divide every three days because they have measured the lifespan of hundreds of these cells. They predict that a red blood cell, born in the bone marrow in January, will die sometime in April because they know that these blood cells live, on average, only 120 days. And, because they have performed dozens of studies on other people, they can guess, with reasonable accuracy, that if you step on a tack, you're going to feel the pain in less than a second.

Still, both the psychologist and the biologist share one philosophy in common. They know that for the human body, as well as for the human mind, timing is everything. The time span of every mental and biological event, the duration of every action has a vital purpose in the survival of the human body. They know, too, that if the timing of any of these events is "faulty" then the body will succumb.

The amniotic fluid that surrounds the embryo and fetus during development is not a stagnant pool. While over 98 percent is water, between 1 and 2 percent of amniotic fluid is made up of substances such as fetal hair and skin cells, enzymes, urea, glucose, hormones, and lipids. It is constantly and completely replaced about every three hours.

Heart as a Time Machine

Take the timing of the heart, for example. Those for whom memories of high school biology are dimming might want to be reminded that the human heart is the fist-sized, four-chambered muscle lying in the middle of the chest that, when it contracts, forces blood to the rest of the body. Its four chambers include two, half-pinky-sized atria residing in its upper part that collect incoming blood from the veins and two larger ventricles in its lower half that squeeze the blood out into the great arteries.

The oxygen-poor blood in the chambers on the right side of the heart never communicates directly with the oxygen-rich blood in its near-mirror image on the left; there is thick muscle tissue separating them. But, each of the atria and ventricles have between them only a flimsy valve that opens to allow blood from the atria to flow into the chambers of the ventricles and then closes when the ventricles contract. It does this so that blood goes out of the heart and not back into the atria.

The heart beats rhythmically. There is a cluster of special dust-speck sized muscle cells, called pacemaker cells, that initiate each heartbeat, setting the basic pace for the heart rate. The pacemaker cells lie within the wall of the right atrium. The cells express their rhythms through periodic fluctuations in their electrical activity; they oscillate between "on" and

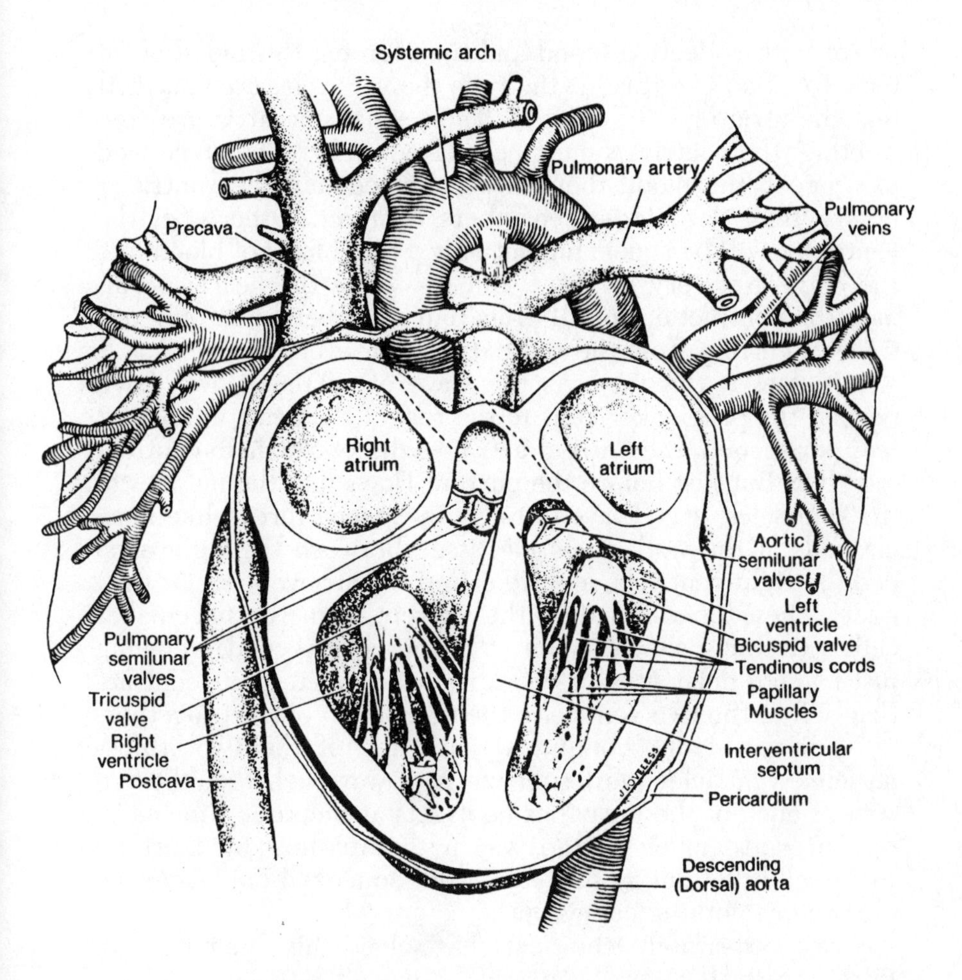

The human heart, the ultimate time machine.

"off." When they are "on," a wave of electrical current flows through them and down through the entire heart muscle. The broad wave of current washes over the entire heart from the right atrium down, exciting the muscle cells and causing them to contract.

But the entire heart does not contract all at once. If it did, it would be very troublesome, indeed. For the heart to do its job of both collecting blood and then forcing this blood out it must function like this: At the time the ventricles are contracting, the atria must be relaxed. Then, when the atria are contracting, the ventricles must be relaxed. Why? The atria need to squeeze the blood they have collected into the ventricles before these two larger chambers contract. Otherwise, the ventricles will be squeezing out only a thimbleful of blood into the body. Not very efficient. But how does the heart coordinate the contractions of all of its four chambers with only one wave of current from the right atrium for every beat?

It does it this way: As the wave of current flows down beyond the pacemaker cells to the rest of the heart at about 2 feet per second, the current activates the two atria to contract together. But just before the current flows into the muscles of the ventricles, it is delayed briefly by special thread-like muscle cells at the wall of muscle tissue between the right atria and the right ventricle, a place called the atrioventricular (AV) node. These fibers conduct the current from the pacemaker cells extremely slowly, about 8 inches per second, allowing just enough delay for the two atria to complete their contraction before the two ventricles then begin to contract together.

The problem, though, is that each of the two, thick-muscled ventricles must squeeze in a way such that the top part of each of these cavities contracts at the same time as its bottom does, much in the way a toothpaste tube must be squeezed to let out its contents (if it's squeezed only from the top, almost nothing comes out).

Not surprisingly, the heart has solved this dilemma, too. Within the wall of cardiac muscle between the two ventricles it harbors specialized cardiac muscle cells called Purkinje fibers. The large cells conduct impulses far faster than any other cardiac muscle cell, about 10 feet per second, and get the current to travel from the AV node to the ventricles in about two tenths of a second. That is fast enough for the cardiac muscles enveloping the ventricles to contract together and for the two chambers to squirt blood into the blood vessels of the lungs and the rest of the body.

The heart is a time machine of sorts that functions by carefully manipulating the squeezing and opening of its four chambers. Its rhythmic "lub-dub" is regular, predictable. The heartbeat is so regular, in fact, that Galileo used it to study the steady motion of pendulums. In 1581, at the age of 17, while kneeling in the Cathedral of Pisa, Galileo observed the swinging of the Great Cathedral lamp and, using his own pulse to measure the time, discovered that the period of the swing of the lamp was always constant, even if the swing gradually died down. "Thousands of times I have observed vibrations, especially in churches, where lamps, suspended by long cords, have been inadvertently set into motion," he wrote afterwards. "But I never dreamed of learning [that each] would employ the same time in passing . . ." Some say that it was from this observation that the pendulum clock was born.

Despite Galileo's keen mind, he was no student of anatomy. Today, physicians and biologists alike know that when the heart tissue is damaged, say, from electrical shock, or from inadequate oxygen to parts of the heart muscle (which often results in a heart attack), then the delicate time machine of the heart becomes chaotic, the contractions, uncoordinated. In a condition known as ventricular fibrillation, for example, the heart beats rapidly, but the contractions are disorganized and confused. Numerous waves of current spread in all directions throughout the heart muscle so that the heart cannot contract as a whole. The result: The heart cannot pump blood into the arteries. The time machine cannot restore its normal rhythm. So the person dies, a victim of a heart with faulty timing.

Syphilis has an incubation period in the body of 10 to 90 days.

Is Timing Sexy?

The beating of the heart is an obvious manifestation of the body's use of timing. But, perhaps more subtle, and infinitely more important than a heartbeat (for the species, anyway), is how the body and mind have so strongly embraced the timing of sex.

As far back as the Middle Ages, sexual timing was on the minds of more than a few Christian theologians. Back then, it was the job of the medieval theologians to interpret the words that St. Paul had written to the people of Corinth: "Because there is so much immorality, let each man have his own wife and each woman her own husband. The husband must give the wife what is due to her, and the wife equally must give the husband his due." Interpreting this text very literally, the theologians put the notion of debt in the center of the sex life of married couples.

Since debt "payment" by one sexual partner to the other was an important issue in St. Paul's doctrine, various moral problems arose in the medieval marriage bed. Was the husband, the theologians asked, bound to prolong copulation until the wife emitted her "semen" (a fluid they believed she secreted during orgasm)? Yes, they agreed, and all of the theologians of the Church of Rome "allowed" the husband to do so. Should husband and wife make their emission simultaneously, they asked. Yes, they said, and all agreed that every effort should be made to achieve this, for simultaneity increased the chances, they believed, of conception and helped to make a more beautiful child. Was the woman bound, they pondered, to emit her "semen" during copulation? History records that of the fifteen members of the Church who took up this question, eight thought that the wife who deliberately abstained from orgasm committed venial sin. The issue passed, thank God.

Whether any of these theologians decided to get the answers to their questions in the bedroom, no one knows. But nowadays, scientists seem to be asking the same kinds of questions about the role of timing in sex but, unlike the theologians, they are getting some fascinating answers. One question concerning sexual timing is being asked by Gordon Gallop and Susan Suarez, two psychologists at the State University of New York.

Humans are bipedal creatures; we walk on two legs. That characteristic allowed our ancestors, at least, to be able to stand up over the tall grasses of the African savanna and survey the surroundings for prey and predators. Bipedalism also had the advantage of freeing the hands for tool-use and for carrying food.

But bipedalism also left us with an important problem. When a female stands, the reproductive tract points down toward the ground. In this position, it is very difficult for the female tract to retain sperm, the necessary ingredient of reproduction. So, they ask, what time mechanisms might the body use to counteract this situation?

Gallop and Suarez speculate that one of the timing gizmos that human females may use to retain sperm is the male semen "plug." In some species of primates, right before the male dismounts the female, he leaves a mucosal mass, or "plug" in her reproductive tract that serves to prevent sperm loss and to prevent other males from inseminating the female.

Human males, though, no longer produce plugs when they ejaculate. Yet, within their semen are mucous particles which are the vestigial remains of what used to be a plug. Oddly, these particles cause the ejaculate to coagulate one minute or so after it is in the vagina and keep it in this gelatinous state for five to fifteen minutes before the particles dissolve. The timing of the formation of this pseudo-plug, they suggest, may aid in keeping the sperm in this reproductive canal long enough for the sperm cells to head for the female egg upstream.

The second timing mechanism to retain sperm within the vagina involves the thrusting behavior of the human male. Unlike humans, monkeys and most apes do not linger over each other when they copulate. On the contrary, the male usually mounts the female from behind, inserts his penis, makes a few quick pelvic thrusts, ejaculates, and separates. Usually, the whole encounter is over in a matter of seconds. Baboons, for example, make up to about 15 pelvic thrusts, lasting roughly 7 to 8 seconds. Rhesus monkeys perform the entire ritual in less than four seconds time.

Fingernails grow about four times faster than toenails—about two one hundredths of an inch per week.

By contrast, humans spend far more time in the act of copulation, usually several minutes more than their hairy relatives. The reason for this is that, unlike monkeys, the human female can experience a sexual climax, but usually takes ten to twenty minutes to do so. But, eventually, the male and, hopefully, the female, reach orgasm.

The timing of orgasm from the standpoint of sexual satisfaction, is quite desirable. But from the standpoint of sperm retention it is vital. According to physiologist Julian Davidson, at Stanford University, orgasm occurs when sexual stimuli from the brain and the genitals pass a critical threshold. When they pass this threshold, nervous signals from deep inside the lower, more primitive portion of the brain, are sent simultaneously in two different directions. One of the signals shoots up to the higher areas of the brain to trigger the intensely pleasurable feelings of orgasm. The other signal fires downward deep into the spinal cord to produce orgasm's physiological reactions in the genital region.

The reason why the timing of orgasm is important for sperm retention is two-fold. During orgasm and ejaculation in males, at least, thrusting of the penis stops when it is at its deepest position within the vagina. This, no doubt, provides the sperm cells with an added edge. The last deep thrust of the penis gets the sperm cells into the closest proximity to the cervical opening, the entrance to the inner reproductive tract and the waiting egg. Any more thrusting would be maladaptive, and a sure sign of a "faulty" timing system. The additional thrusting would only serve to draw the semen back out of the vagina. Luckily, though, the female orgasm timing mechanism sets up a series of contractions in the muscles surrounding the outer third of the vagina which produce a stopper-like effect and hold sperm in.

Hair, unlike nails, does not grow continuously; it is lost and renewed periodically. Scalp hair grows about one inch every two or three months. Coarse hair grows about half as fast as finer ones. Cutting or shaving hair has no effect on its growth.

Far more interesting than the primitive motions occurring down in the pelvis, are the psychological reactions that occur during and after orgasm. For both in the male and in the female, intercourse frequently ends by producing feelings of sexual satiety and a mild sedation effect that may last anywhere from three to seven minutes.

The timing of this brief repose is important for the female.

It is designed to delay her rise to an upright position until the sperm are safely deep within the reproductive tract. Indeed, some scientists have suggested that the multiple female orgasms themselves may function to enhance the probability that the female will remain supine when the sperm cells are still swimming about in the vagina.

Timing, Touching, and Gazing

In the human species copulation is usually not possible without courtship. And courtship involves some intricate timing mechanisms of its own. The duration of touching, of course, plays a large role in courtship. At first, the touch is a brief pat on the hand or a quick hand-shake between two strangers. But, if the two then find some attraction with each other, they will hold hands for a longer period of time. Gradually, the touching becomes prolonged and more intimate. Eventually, it becomes part of a loving embrace, and then a sexual one.

Social eye contact plays a role in courtship, as well. The earliest form of this contact begins when the male (female) looks at another female (male) with a series of quick glances. In a fraction of a second, though, all of the physical and often emotional characteristics of that individual are summed up and mulled over. And, if all of the signs reveal that the person of the opposite sex is attractive, then the next sequence occurs.

At first, eye contact is brief. The usual reaction is to look away quickly, to break eye contact, to avoid that temporary invasion of privacy. Indeed, if one member *does* stare for too long at the other, the other party will become acutely aware that they are being ogled and will move away from the probing eyes. At this early point in the relationship, the lengthy stare is in itself an act of aggression between unfamiliar adults. So rather than meet each other's eyes, they watch one another in turn, looking away when they spy the other watching. Eventually, though, their gaze will become prolonged and more intimate, and they will smile.

And, predictably, even *that* will have an element of time to it. A smile will begin when the mouth widens and the corners of the mouth pulls up. Then, as the upper lip is raised, it will partially expose the teeth and also bring about a downward curving of the furrows which extend from both nostrils to the corners of the mouth. This, in turn, will produce a puffing, or rounding out of the cheeks on the outer side of the furrows. Creases will occur momentarily beside the eyes (in older people, it is possible to see "laugh lines" at the edges and below the eyes). Moreover, the eyes will themselves change, becoming "bright and sparkling." The twinkling of the eyes will be caused by a small amount of moisture produced by the tear glands.

Time Clock in the Brain?

The length of a gaze, the duration of a touch, the interval of a smile, and every other expression can be timed and measured and, indeed, have been. That alone must mean that we have some mental awareness of the passage of time of these social signals since we are able to mentally manipulate their durations. And *that*, must mean that we have some internal clock ticking in our brains that measures the time.

So far, though, no one has found the "clock" in the brain that is responsible for our ability to measure the passage of time. The hunt for such a clock has led to a feeling of frustration among some investigators. "As far as we know," wrote Alfred Kristofferson, a leading psychologist at McMaster University in Ontario, "the system with which we deal may contain any number of clocks, including none at all."

In rats, at least, scientists have selectively removed one section of the brain then tested the animals' ability to measure the time intervals between two successive sounds. Then they have removed other sections of the brain. To date they have found nothing to indicate that a clock that measures mental time resides in the brain. Some researchers suspect that if there is a clock, it may reside in every cell of the body.

Still, there are a lot of professional comedians, athletes, photographers, and grenade throwers who depend on the accuracy of this as-yet-to-be-discovered clock for their livelihood. To them, a good sense of timing is a gift. A bad sense of timing is a curse. ("Bad timing in this business," said one comic, "is like bad breath. Sooner or later, they're going to find out you stink.") Jack Benny was an expert at squeezing out a laugh from a well-timed pause. There is, of course, the classic sketch about his well-known miserliness. In the scene, a gunman demands of him, "Your money or your life," Benny looks at the audience. After a long pause—just long enough to elicit some giggles from the audience, the impatient gunman demands an answer. "Well?" he asks. Replies Benny: "I'm thinking, I'm thinking!"

Why Time Flies and Drags

If scientists do manage to locate this "clock" in our brains, they may find out what others have suspected all along, that the clock is responsible for our pre-occupation with variable time—time dragging or flying. Indeed, a certain span of time can seem to pass slowly or quickly depending on the emotion. In darkness and waiting or during anxious anticipation, minutes and hours seem to last much longer; in gaiety and joy, time passes too quickly.

In 1932 physiologist Hudson Hoagland began to ask why time tended to fly or drag when his wife was ill with the flu. She had been lying in bed with a temperature of 104 and had asked her husband to run to the drugstore to fetch some supplies. When he returned, she was angry. "Although I was gone for only twenty minutes," he wrote of the incident, "she insisted that I must have been away much longer." Hoagland was stunned, and fascinated. "Since she is a patient lady, this immediately set me to thinking."

As a scientist, Dr. Hoagland had been reading about the work of the French psychologist Henri Pieron who, in 1923, suggested that if the speed of our physiological processes were modified by, say, temperature, the result would be a proportionate increase or decrease in our mental time. Pieron's suggestion inspired the dramatic experiments of his pupil, Francois. In one situation, Francois asked his students to tap a key at the rate of three times a second. He then raised their temperature with heating pads wrapped around their bodies and found that the rate of their tapping increased. Pieron argued that just as clock time seems to pass slowly when body temperature is high, so clock time would seem to pass rapidly when temperature is lowered.

The reason for this is that heat speeds up chemical reactions. Indeed, the beating of the hearts of cockroaches and the frequency of chirping of crickets, which depend on chemical reactions, increases as a function of temperature.

If time is determined by chemical velocities, then raising our body temperatures should speed up the reactions. The increased reactions should make more chemical changes and hence make physiological time pass more quickly in a given interval of clock time than would normally be the case. Lowering the internal body temperature should have an opposite effect, making clock time seem to pass faster. In a fever, Hoagland argued, we should show up early for appointments, and with lower body temperatures, we should show up later.

The inner lining of large blood vessels renews itself every six months. The skin is even faster, renewing itself every three to four weeks. But, by far the fastest renewal rate takes place in the inner lining of the digestive tract. Both the cells of the stomach and the entire intestinal lining are replaced every three days.

Now, Hoagland believed, he had discovered the same phenomenon occurring somewhere in his wife's brain. He frantically searched for a stopwatch while she lay there in bed. Returning to her bedroom, he asked the sick Mrs. Hoagland to count out loud to sixty "at a speed she believed to be one per second." Days later, when she was well, Hoagland tested her again and was delighted at the results. He had discovered that his wife counted faster when she had a fever than when she did not.

That our sense of time would depend entirely on the rates of biological reactions in the brain might, at first, seem a bit far-fetched. But it is interesting that certain disease states, like hyperthyroidism, characterized by an over-production of thyroid hormones, which increase the metabolic processes of cells, can alter the perception of time, also.

Certain fever-producing drugs such as mescaline, also increase metabolic rate. Mescaline not only raises body temperature but also produces an overestimation of clock time. Handwriting samples of volunteers on the drug are sloppy and spread out. And people on psilocybin, also a pyretogenic or fever-producing drug, actually speed up their tapping rate showing that time passes, to them, more slowly.

And maybe it's just a wive's tale, but there is a feeling most people get when they reach middle age that the years go by faster than they did when they were children, when summer seemed to last centuries. Anthropologist Edward Hall suggested in his book, *The Dance of Life,* that one of the reasons for this may be simple mathematics. To a five-year old, a year represents 20 percent of his life. To a fifty-year old, a year makes up only 2 percent. And since we live life as a whole, Hall argues, our perception about the duration of a year changes with time.

But, some biologists have another explanation. Back in the 1950s, biologist Semour Kety found that there was a rapid fall in both the circulation and oxygen consumption of the brain from childhood through adolescence. This was followed by a slower decline towards old age. According to his theory, the slackening of brain oxygen use with the slowing of cellular metabolism with age would tend to make time appear to pass more rapidly as we get older.

Brain Time

In 1982, physiologist Benjamin Libit of the University of California was not looking into how we feel about time, but about how the timing of events occurs in the brain. Libit wanted to see what was happening in the brain at the time one decides on some motor act. Libit showed that there is a fraction of a second of time between one's conscious awareness of an impending action, like wishing to flex one's fingers, and its actual occurrence (flexing the fingers).

According to Libit, unconscious actions always precede the actual movement of a finger or a limb. In one experiment, Libit asked six student volunteers to flex a wrist or finger any time they felt like doing so. Earlier he had placed electrodes on the subjects' heads so that he could monitor their "readiness potential," a sudden change in electrical activity generated by the brain just before an event is about to occur.

Among his six subjects, these electrical potentials began about eight tenths of a second before a muscle flexed, while the students were consciously aware of their intention to move about three-tenths of a second later.

Libit believes that the split-second delay between the awareness of an intention and the actual movement of the muscle provides an opportunity for people to consciously control their actions. When Libit asked his subjects to first think about moving their muscles and then to see if they could consciously block the limb movements a split second before they budged, the "readiness potentials" appeared about a second before the time they were supposed to move their limbs, but the electrical activity of the brain dampened around two-tenths of a second before movement would have taken place.

Somehow the brain uses this conscious mental veto power to interfere with the final development of these potentials. This may explain how we are able to have rapid self-control over our urges to act.

Libit would readily admit that speedy thoughts are common currency in the human brain. But by the same token he would also confess that, in the broader sense, the speed is relative. According to a number of researchers, including Yale University psychologist, Robert Sternberg, bright people have, in general, quicker minds than less intelligent individuals. They are, for the most part, far more rapid processors of information than their less bright neighbors. They retrieve information, and probably store it, more rapidly than those individuals with duller minds. They breeze through the Scholastic Aptitude Test (the SAT), acquire lengthy vocabularies the way most people acquire lint, and in school, they are quicker on the draw with the right answer than their classmates. How do they do it?

Psychologists like Sternberg suspect that these gifted individuals actually spend much of their time encoding information to assure that they have processed the information richly and in detail. They therefore avoid the situation later on in which they have to change the way they encoded this information. Encoding involves translating a word like *love,* or *France,* into a mental representation.

Take the verbal analogy problem: WASHINGTON is to ONE as LINCOLN is to (a) FIVE, (b) TEN, (c) FIFTEEN, (d)

TWENTY. To begin to answer this problem, most people might encode the information that Washington was a president (the first). But a bright individual would encode the word WASHINGTON even further. She would note that Washington's face is also on a bank note, and that he was a war leader. The additional (albeit brief) time encoding would lead the bright individual to quickly associate Washington with the $1 bill and therefore Lincoln with the $5 bill (and the correct answer, FIVE). Her more ordinary classmates, having limited their encoding of the word WASHINGTON, would be left behind, still trying to figure out whether or not Lincoln was the tenth, fifteenth or the twentieth President of the United States (he was the sixteenth, a choice not included in the exam).

In the testes, a thousand sperm cells are completed every second (it takes about two months to manufacture a fully mature one). After ejaculation, sperm head for the ovum, or egg cell, propelling themselves at a speed of about 4.2 inches an hour (the equivalent of a swimmer covering about 12 yards a second) and reach the site of fertilization within about fifty minutes. There, sperm can remain alive for roughly two days.

Almost by definition, information processing and retrieval mean that changes have to take place in the brain, and that at least some of these changes must be stored in nerve cells. But how these organic events occur is still largely a mystery. No one even knows how nervous connections are established or rearranged during learning. The nervous system of humans is extremely complex. But neuroscientists suspect that the neural circuitry of exceptionally gifted people is different from the neural connections in our brains. How else would one explain the abilities of an Albert Einstein or a Truman Henry Safford?

Born in Royalton, Vermont, on January 6, 1836, Safford was a child prodigy with a reputation for being a "lightning calculator." When a child, his friends would take him out into the countryside and ask him how many palings there were in the long fence *over there.* He would sweep his eyes over the entire fence, and then say without delay 147, and his friends would go over and count each stick in the fence and find exactly 147 of them. Then they would find another fence, and ask him how many sticks there were in *that one*, and he would scan the fence briefly and tell them 247, and they would count them and discover that he was right.

When Safford was 10 years old, a Reverend Henry Adams, who had heard of the boy's amazing powers of memory and mental speed, asked him to multiply in his head the number 365,365,365,365,365,365 by 365,365,365,365,365,365. "He flew around the room like a top," wrote Adams, "pulled his pantaloons over his boots, bit his hand, rolled his eyes in their sockets, sometimes smiling and talking and then seeming to be in agony, until not more than one minute, said he: '133,491,350,208,566,925,016,658,299,941,583,225.' . . . [Afterward] the boy looked pale and said he was tired. He said it was the largest sum he ever did."

Safford died in 1901 at the age of 65, but not before preparing almanacs for Boston and Philadelphia (at the age of eleven), completing college in two years and graduating with honors from Harvard (at the age of eighteen), and discovering several celestial nebulae (because he remembered the positions of all the stars in the Nautical Almanac and therefore knew what was out of place).

Safford's mental speed was exceptional and still difficult to explain. Some neurobiologists suspect that mental speed may be accomplished when certain key neural circuits are "hardwired" in the brain. The neural "hardwiring" may help to accelerate commands. Other researchers suggest that gifted individuals possess more energetic brains than ordinary people. Like the spinning tires of a race car at the starting line, most of the nerves in their brains may always be energized but held back, and it may only take the release of a small number of neural "brakes" to let their signals go.

The same processes may also occur in the brains of professional athletes, who seem to exhibit a vast array of speedy movements; who hit, dive for, swing at, and catch a ball faster and more accurately than ordinary folks. But these are just guesses. No one really knows.

Still, those of us cursed with bad timing have always been quietly respectful of those who possess both a quick mind and split-second timing. The British statesman Winston Churchill had this talent. The story goes that one day when Churchill was delivering a speech on the floor of the House of Commons, an angry Nancy Astor, the first woman elected to this body and a member of the "bad timing" group, stood up and interrupted him. "Winston," she shouted, "if you were my husband, I would flavor your coffee with poison!" Churchill looked at his mental underling for but an instant: "Madam," he said, "if I were your husband, I should drink it!"

TWO

THE BODY IN MILLISECONDS

G. Gordon Liddy performed a neat trick at parties. The ex-CIA man and former Watergate prankster would hold the palm of his hand over a candle's flame until his flesh burned. While the guests gawked, Liddy would stand there motionless as the flame licked his hand. Then, after he was satisfied that his machismo had been observed by all, he would remove it. The pungent smell of scorched skin still permeated the room after he had left.

Liddy was no biology professor, but he had a good science teacher's knack for demonstrating how the normal human body is supposed to work. Unless we actively or consciously inhibit it (which was what Liddy did), the body has its own way of protecting itself from harm. It's called a reflex. And it's awfully fast.

All human reflexes are unlearned, involuntary, and entirely predictable. Designed to defend or buffer the body from a mercurial and often harmful environment, they occur in response to sensory stimuli.

The seventeenth-century French philosopher and mathematician Rene Descartes, was one of the first to describe the reflex. Colored by the thinking of the time, he believed that animal spirits from the brain, the seat of the human soul, were reflected (hence the term reflex) into the muscles of the body causing them to twitch.

Nerve regeneration takes on average four to six weeks.

If Descartes' notions of the human body were appealingly romantic, they were nonetheless completely wrong. Reflexes involve simple nerve circuits mostly between muscles and the spinal cord, although the brain can be involved, too. Though it never wins much attention at parties, the sudden withdrawal of the hand from a flame is a reflex. Sensory receptors in the skin transmit nerve impulses to the spinal cord. The impulses then fire back through special motor nerve cells to the muscles which control the movement of the hand. The muscles contract; the hand jerks away. A "split second" later, we feel the pain. Yet, the hand is saved.

Reflexes come in all forms. There are reflexes involved in adjusting the diameter of blood vessels, reflexes which affect the activity of bowel movements and breathing. Some are complex, involving many nerve connections; others are not. What they have in common is their speed. Three of the most familiar

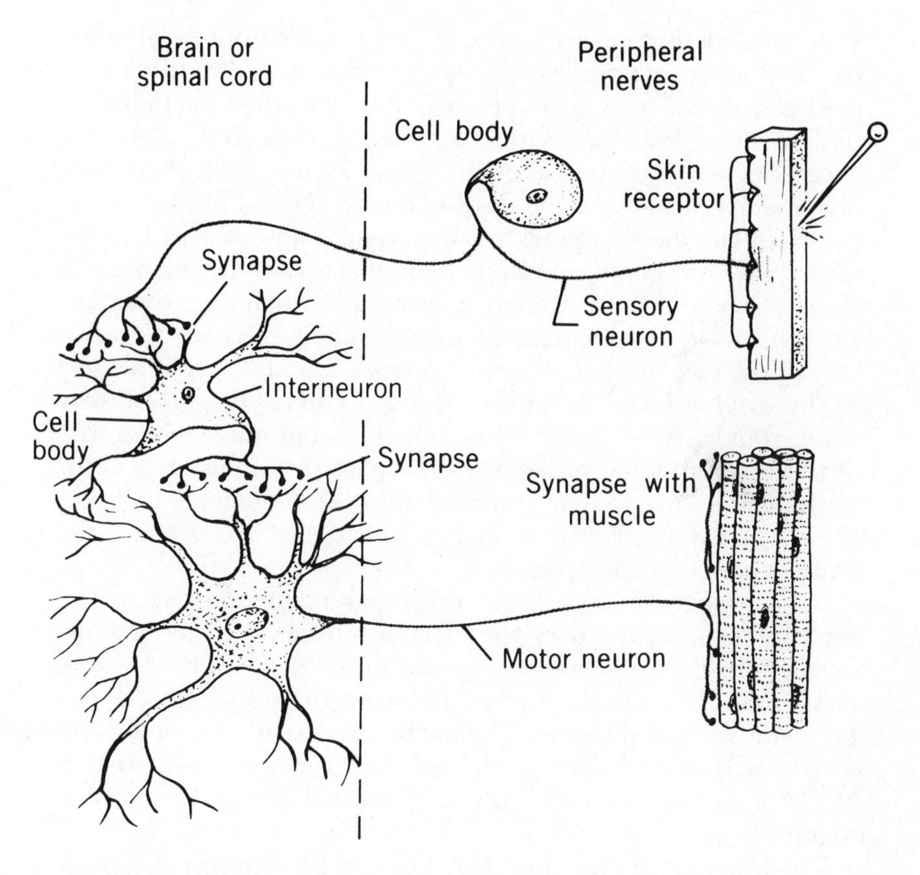

A reflex is fingersnap fast and usually involves a simple circuit of nerves and muscles.

reflexes—sneezes, hiccups, and eyeblinks—are also three of the fastest. How come?

Eyeblinks are vitally important to the body, for they preserve and protect the eyes. In times past, the eyelids were sometimes snipped off as a method of torture. Victims of such cruel punishment eventually went blind as the cornea, or transparent coat of the eye, dried up and clouded over.

There are really three different types of blinks. The reflex blink is a protective response. It is triggered by anything from a loud and sudden noise or a tap on the forehead, to a baseball suddenly coming out of nowhere. We can also blink voluntarily. Voluntary eyeblinks last longer than a reflex blink and are used consciously to wipe away any grit, dust particles, or pollen grains that have fallen onto the cornea. But, there are spontaneous eyeblinks, as well. These blinks have their own internal cycle, and are the most common type of blink.

Spontaneous eyeblinks wash, spread, and mix tear fluid—a complex film made up of oils (which act as lubricants for the sliding action of the lids), sugar-proteins (which act as wetting agents), salts, and bacteria-digesting enzymes—over the outside of the eyeball, keeping the cornea clear. The eyeblink acts as the driving force for pulling, by a suction action, fresh tear fluid from the tear glands (fluid which is continuously secreted at a rate of about 1 millionth of a quart per minute), and for pushing the "used" tear fluid out into the tear ducts and into the nasal cavity. Tears drain into the throat even when the body is hanging upsidedown.

The upper lid takes on the role of a kind of miniature windshield wiper, sliding over the exposed cornea and shoving any debris into the space along the margin of the lower lid where it is washed away. This contact between the upper lid and the eye is intimate and strong. The eyeball is actually pushed down a sixteenth of an inch during each blink, a movement that is so fast it can be observed only with the aid of high-speed photography.

While one of the chief functions of blinking is to protect the eye surface, researchers have recently found evidence to suggest that blinking serves a role far beyond cleansing of the eye. If the eyes are the windows to the soul, the eyelids, placed as they are at the portal entry into the eyes, are the screens reflecting the activity of the brain itself.

Humans blink, on average, about 24 times a minute (dogs and cats, in comparison, blink about 2 times a minute). This rate, though, is deceptive. The frequency at which we blink changes with our moods. When we get bored or tired, we blink more frequently. Driving in a car, for example, we may at first blink on average about 15 times per minute. The duration of each blink could be actually only 200 milliseconds, or two-tenths of a second. An hour later, while driving on a long stretch of highway, we may blink at a higher frequency, about

40 times per minute and the duration of each blink could last about three times as long as it did when the trip first began.

Humans blink more frequently when they are angry or when they are talking to strangers or to the boss than when they are just chatting with friends. In court, witnesses under cross-examination blink more frequently than they do at other times. Emotional blinking in response to embarrassing situations or questions, researchers suggest, may be defensive or submissive—a momentary withdrawal of attention from the cause present in the immediate surroundings.

Blinking occurs during a change of gaze and often initiates an eye movement. Reading significantly decreases blinking (although we often blink each time we come to a period at the end of a sentence), the more difficult the text, the less frequently we blink. When we try to consciously suppress blinking for a few minutes (four is usually the limit), we blink even more frequently immediately after to make up for the loss—a phenomenon known as the rebound effect.

In 1981, Craig Karson and his associates at the National Institute of Mental Health in Washington, D.C., had their subjects read, or sit in silence chewing gum. They found that speaking or reciting a proverb by memory increased the rate of blinking in their subjects. However, when the students simply interpreted a proverb, they blinked less. Gum chewing had no effect. Apparently, the mental stress of reading from memory, and not the movements of the mouth, affected blinking rate. They also found that men were able to consciously speed up their blinking at a rate faster than women, though the scientists didn't understand why these differences occurred.

Arterioles, or small arteries, contract and then relax every two to eight seconds.

That blinking reflects the function of the brain is even more obvious when one looks at the maturation of blinking in children. Very young infants, up to two months of age, blink slightly less than once a minute; by the age of five to ten years old, children blink at the rate of 6 times a minute. The rate of eye blinking accelerates after that reaching its peak of 24 times a minute at the age of 20. It stays at that rate throughout life and into old age. As the young don't get any less dust or grit in their eyes than older folks, blinking must serve a purpose that is more than just eye protection.

In the past, eyeblink researchers believed that the upper eyelid, like the shutter of a camera, was the principle blink-producing mechanism. It's not. Actually, the movements of both the upper and the lower lids are involved in the human eyeblink and form a complex and interesting pattern.

In 1964, David Kennard and Gilbert Glaser working at Yale University, spent months analyzing the different parts of an eyeblink and found it to be quite complicated. Using a system of levers attached to the eyelids of their human subjects—a group of undergraduate students—they described a number of phases in the dynamics of the eyeblink that had not been studied before.

Kennard and Glaser showed that there are not one but three distinct phases in the initial closing portion of the blink, that part of the blink when the lids close down over the eyes. The movement of the closing portion of the blink is akin to driving a car. To get the car going, we first have to accelerate from zero to our cruising speed. Then, once at our cruising speed, we have to apply the brakes to slow down.

Like a car, the eyelids begin their journey over the eye from a dead start. For the first 10–20 milliseconds, the lids simply accelerate from their once open resting position. This is the first, slow phase of a blink. Now up to their cruising speed, they move rapidly towards the middle of the eye. This phase, called the fast phase, lasts from 20–50 milliseconds. The third phase, which lasts from 20–50 milliseconds, occurs when the lids decelerate on their way toward each other near the middle of the eye. The journey ends when the two eyelids meet and the eye is closed.

This, though, is really only half of the story. During the initial closing phase of a blink, the lower lid moves first, preceding the movement of the upper lid by approximately twenty milliseconds. Why this happens is not entirely known. One suggestion is that the lower lid is lighter than the upper one and so moves faster. Another suggestion is that the lower lid moves faster because the eyeball cornea is smoother beneath it and so the frictional forces between the lower lid and the eyeball are less than those between the eye and the upper lid. Still a third observation is that the nerve running to the lower eyelid is shorter than the one going to the upper eyelid. The nerve signals from the brain to the two lids may leave at the same time, but the signal gets to the lower lid a few thousandths of a second faster than it does to the upper one. So, the lower lid rises sooner. Whatever the dynamics, the time from the initiation of lid movement to full eye closure is short, generally lasting from 50–145 milliseconds.

When the two lids come smashing together near the middle portion of the eye completely covering the pupil of the eye, they remain fused for only a brief period of time, approximately 50 milliseconds. This brief duration, though, increases when we become drowsy, to about 80 milliseconds.

After measuring the closing portion of the blink, Kennard and Glaser then determined that the reopening portion of a blink was as complex as the closing sequence. It had two phases. First, the lids peel off each other and fling back quickly. This is followed by a progressively slower motion as the lids decelerate and settle into their fully opened position. Reopening time for the blink was relatively long with the full combined reopening phases lasting from 100–200 milliseconds. This, they found, was almost three times as slow as the opening phase of the eyeblink. For a reflex, this makes sense. It is more important to raise a shield quickly against a projectile, than it is to lower it once the projectile is gone.

Recently, though, scientists have been looking at the electrical activity of the brain while a blink occurs. Doing so, they have found one of the most bizarre aspects of blink activity discovered to date. One obvious fact is that while the lids are closed over the eyes, we cannot see. What is not obvious, though, is that as much as 50 milliseconds *before* the lids begin to leave their homes, the visual system of the brain shuts down. Why? Supposedly, at the same time the brain fires off a signal to the muscles of the lids, it also fires off one to the visual region of the brain. This signal tells that region of the brain to shut down. In effect it is saying, "there won't be any information coming to you for the next four tenths of a second or so. Don't bother processing."

Whatever the reason for this system shutdown, this event, coupled to the actual blink, which takes not less than 200 msec (and averages about 400 milliseconds), produces a significant period around each blink during which vision is either blocked or reduced. What is surprising, though, is that we are not aware of this gap in our vision even though we can perceive even briefer durations when the room lights are shut off then on again.

The amount of blood passing through the arteries feeding the heart (the coronary arteries) is approximately two tenths of a quart per minute or about four to five percent of all the blood pumped by the heart.

Learning about the durations of all the separate phases of an eyeblink might appear esoteric at first. It might . . . until one looks at the agencies sponsoring such research: the United States Air Force, for one. Their interest is obvious. If a tired fighter pilot blinks at the rate of 40 times every minute, and if the lids of an individual eyeblink blot out vision for roughly three tenths of a second (including the time during which the brain's visual system is itself shut down), then a pilot would be oblivious to the environment for about 12 seconds out of every minute of flight time. Going two or three times the speed of sound, a pilot could travel some five miles with his eyes closed. Incidentally, the commercial pilots for the passenger airlines get tired, too.

Understanding the speed of a normal eyeblink and the rates at which it occurs in healthy subjects, is important for another reason, as well. For those in psychiatry and in the neurosciences, knowing the dynamics of a normal eyeblink makes it a precise and useful tool for studying certain diseases of the brain. Blink rates have been found to be higher than normal, for example, in patients with schizophrenia, Gilles de la Tourette's syndrome (where patients display muscle spasms and tics), and tardive dyskinesia—illnesses in which the brain chemical dopamine is very active. On the other hand, blink frequencies are lower than normal in those individuals who suffer Parkinson's disease and other parkinsonian syndromes. Interestingly, the latter possess a lower level of dopamine in their brains. Studies such as these suggest that brain dopamine plays a major role in normal spontaneous blinking.

The brain controls the rate of blinking, and some researchers have discovered that blinking itself can activate the brain. Recent studies have shown that blinking transforms light into pulses onto the retina. These light pulses then activate certain nerve cells in the brain which respond specifically to these flashes. The nerves, in turn, fire off signals of their own to other regions of the brain.

A burst of blinking can be a good thing, protecting our eyes from a cloud of car exhaust or the smack of sunlight. For some, suffering from the ailment known as blepharospasm, the frequency and force of blinking are so intense and fuse into such fluttering spasms, they often become serious enough to make one functionally blind. Usually it occurs without warning, often associated with some serious brain disease. But it can also be inherited, appearing most frequently during early childhood. For those with this troublesome problem, it can be psychologically devastating, as well. Sufferers stop reading, stop watching television, and avoid social engagements.

For most of us, though, a blink of the eye is but a transitory event, coming as it does, every so often. Even so, we spend most of our waking lives oblivious to its actions. Perhaps its speed is the reason that we have for so long paid it no heed. Today, though, the speedy eyeblink is more than just a queer event.

The eye, exposed as it is to the surroundings, needed a way of protecting itself. But the eye is not the only organ of the body naked to the world. The lungs are too. Air—filled with bacteria, various viruses, dust, dirt, fibers, and insect droppings—is sucked into these two gas exchange sacs every few seconds. Luckily for us, most of these particles are trapped in the nose or along the sticky, mucous lining of the throat; their fate left to slide into an acid stomach or to be blown out into a waiting Kleenex. But the lungs are protected by a powerful and speedy reflex, as well: the sneeze.

A sneeze is a strange creature. Fingersnap quick, often

firecracker loud, and embarrassingly imposing, its nature was, and in part still is, mysterious. A protective reflex, it is mostly triggered by irritants in the nose. Yet, most of the blast of air, to the displeasure of anyone standing in front of it, comes out of the mouth. One of the major defensive reflexes of most mammals, it is selective in humans. Caucasians sneeze more often than Blacks; males more than females. Moreover, lodged in the sneeze is another protective reflex—the automatic closing of the eyes. Even the cough, its closest cousin, occurs with eyes open.

The lifespan of a blood platelet is only ten days.

Throughout the ages, the sneeze has been the subject of discussion and superstition. The established Church once thought its purpose was to clean out the brain. Even Dr. Oliver Wendell Holmes, the nineteenth-century American poet, humorist, and physician, felt that it was healthy for newborns to sneeze.

According to Greek mythology, Prometheus first introduced the sneeze to mortals. The story goes that he built a statue made from stone, then changed his mind and decided to give it life. To do this, he stole some sunlight from Apollo's chambers and hid it in his snuffbox. A few days later, when his mind was elsewhere, he took a pinch of snuff ala sunbeam, put it up his nose, and sneezed.

Some ancient Greek scientists had their own opinions about the sneeze. Hippocrates considered sneezing dangerous in diseases of the lungs. However, he also suggested that sneezing might cure the hiccups. Sometimes it does.

The common practice of saying "God bless you," when anyone sneezes is almost universal. Most historians, though, believe that it probably originated with Pope Gregory the Great (540–604 A.D.) during a major plague. He introduced the short benediction to be used only when someone sneezed, because sneezing was associated with death. Soon thereafter, when someone sneezed, people said, "God bless you," but they meant, "God help you." During the Middle Ages, when another plague raged, an English children's nursery rhyme emerged:

> *Ring around the rosy (the ring rash)*
> *Pocket full of posy, (a bunch of flowers)*
> *Achew! Achew!*
> *All fall down (all fall dead)*

Perhaps because of its history, or because it occurs so quickly (and so violently), a sneeze had, for some time, been

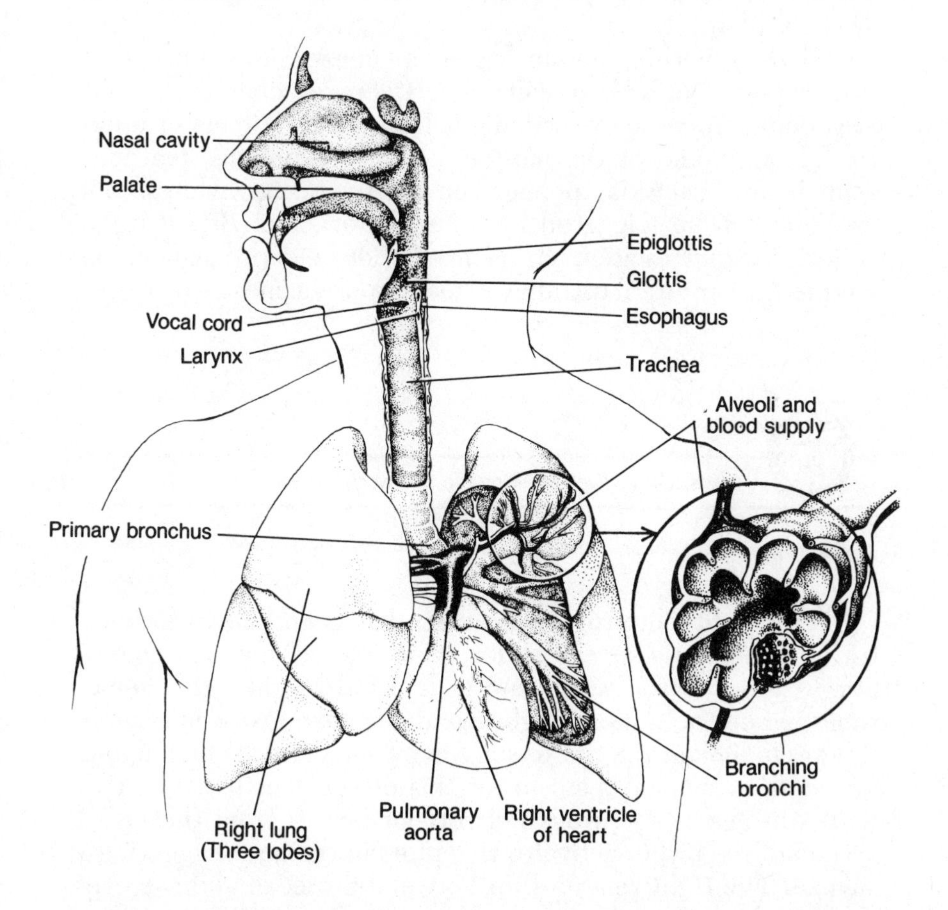

A sneeze is a strange creature that involves the lungs, the windpipe (or trachea), and the glottis.

difficult to measure. With the advent of techniques such as X-ray motion pictures, by which one can view the movements of body parts, and special devices that can record air flow going in and out of the lungs, the events in a sneeze are no longer as mysterious as they once were.

To many, a sneeze *is* just the way it sounds. Actually, though, the *Achew!* is only one phase of a sneeze, the one researchers call the respiratory phase. This mostly extroverted part is simply the result of air being drawn into the lungs (the *Aaaaaaah* of the sneeze), then rapidly released into the surroundings in one wet and powerfully explosive . . . *chooo!* But sounds can be deceiving, and a sneeze is more complicated than it sounds.

First, something has to trigger a sneeze into action. That "something," may be a wide assortment of characters. The most common cause of sneezing is from local irritants or allergic factors. Some of the most effective are pollens, feathers, animal hair, bacteria, pepper, or for some, egg whites. But sneezing can be triggered by other factors, too. Bright light, chilling, sexual excitement, menstruation and pregnancy, resentment, fear and frustration—may arouse a sneeze or two.

Gums are renewed every one to two weeks.

That light can provoke a sneeze has been known for centuries. The photic sneeze reflex, as it's called, occurs immediately after a person walks into or looks at bright light. Sometimes, when the nose is tingling and a sneeze just won't come, it helps to look at the sun or stare at a room light. While no one knows the exact mechanism for this effect, it is believed that light, through a torturous network of nerves from the visual system of the brain, activates receptors in the nose triggering a sneeze. The tight relationship between a sneeze and the eyes can be demonstrated by its behavior: Bright light triggers blinking and tearing reflexes. So does a sneeze.

Normally, only about 20 percent of the population sneeze each time they walk into bright light. Some researchers have suggested that this "sneezer trait" is inherited and is passed on to offspring in a dominant fashion, the way brown eyes are passed onto the kids when mom has brown eyes and dad has blue. Researchers even have a term for this genetic trait. They call it: ACHOO (for *a*utosomal dominant *c*ompelling *h*elio-*o*phthalmic *o*utburst) syndrome. Why should this trait be dominant? There is a selective advantage to being an overly-sensitive sneezer. In arctic and other cold climate populations habitually exposed to respiratory infections, frequent sneezers get sick less often. They expel germs before they have a chance to get into the body.

Whatever the cause, once a sneeze is activated, the only thing left to do is to helplessly wait for it to run its course. This is how a sneeze runs its course:

Usually the first event is when a cat's hair or some pollen floating in the air is suddenly sucked into the nostrils. For a brief moment, nothing happens. Then, unbeknownst to us, their presence triggers special cells lining the inner part of the nose which immediately send their nerve impulses to the medulla situated at the base of the brain. So far, only about three dozen or so milliseconds have gone by.

Provoked into action, the medulla fires its own signals back to the nose, this time routing them to a special nerve cluster lying near the lining of the nostrils. A few milliseconds later, impulses from this nerve ganglion stimulate the mucous glands in the nose to secrete a clear though slightly viscid fluid. They also cause the tiny blood vessels embedded in the soft membrane of the nose to swell.

At the same time, nerves from the nose fire their battery of impulses to a sensory region in the brain. Gradually, we begin to feel a peculiarly pleasurable tingling or quivering feeling, caused by the secretion of mucous in the nose. This feeling usually lasts anywhere between 2 and 15 seconds. The length of time depends on the amount or even the type of irritant in the nose and the speed with which the nose produces its mucous secretions. Eventually, the tingling, tickling feeling gets so intense that we yearn to sneeze.

To speed the process along, we stare at the ceiling light. This causes the inner lining of the nose to engorge even more than before. The congestion and secretions of mucous in the nose irritate the trigeminal nerve and the fibers of the nerve fire a rapid volley of impulses back to the medulla.

Aaaaah . . .

Aroused once more into action, the brain's medulla accelerates its slow stream of firing up to a cool 300–400 pulses per second. The nervous signals travel to the muscles of the lungs. They contract. For 4 to 7 seconds we suck in air, filling our lungs with about two and a half quarts of the gas. Inside the lungs, air pressure rises. The balloon inflates. Then, *click*. The glottis, or upper airway, slams shut and the vocal cords close tightly.

For two hundred milliseconds, the air within the lungs remains trapped. But events are still occurring. Within this splice of time, the medulla switches loyalties. Signals that had called the inspiratory muscles into action begin to slow down. The lungs stop trying to suck in air.

The medulla now fires a battery of pulses to the muscles of expiration. The muscles around the abdomen and the ribs contract forcefully, pushing against the diaphragm muscle and against the rib cage. Still trapped, the air within the lungs is squeezed. Air pressure rises to dangerously high levels. The balloon is about to burst. The two tenths of a second finish. And. . . .

chooooo!

The vocal cords and the glottis suddenly pop open. The back of the tongue rises. And two and a half quarts of air explode outward, expelled by the rapid contraction of the muscles of the abdomen and ribs. Air blasts out of the lungs within a half second. Automatically, the lids slam shut over the eyes protecting them from the potentially infectious air, and the small muscles in the middle ear contract, protecting the delicate bones there from the noise. The warm, moist air blasts into the environment carrying with it any debris, pollen, or bacteria that happened to have attached themselves to the lining of the windpipe.

How fast does the air move through the windpipe during a sneeze? In 1955, B. B. Ross and his associates at the University of Rochester in New York asked that same question and came up with a fascinating discovery. Looking into the mechanism of the cough, which goes through the same basic movements as a sneeze and performs roughly the same function (the sneeze has been described as an upper respiratory cough), Ross found that the air from the lungs is expelled at close to the speed of sound. Later experiments showed that the same thing happens during a sneeze. How is that possible?

During a normal gentle exhale (of the type we do all the time), air travels through the windpipe at a speed of roughly 15 miles per hour, barely a breeze. If, however, the air in the lungs is squeezed out forcefully by contracting the muscles of the abdomen and the ribs, the speed at which air flows through the windpipe increases to a velocity equivalent to 100 miles per hour. These are hurricane force winds to be sure, but it still does not explain the much higher rates coming out of a sneeze.

Ross and his associates uncovered the answer when they examined the high-speed X-ray photographs taken during an active cough. During a cough, as in a sneeze, the hollow windpipe collapses like a rubber inner tube just before the air leaves the lungs. When the glottis opens, the air is literally squeezed out of the lungs and through a collapsed windpipe in just the way water under pressure is blown out of a small diameter hose. Air travels out through the windpipe at nearly 85 percent of the speed of sound, a velocity that is more than enough to tear away any particle or bacterium foolish enough to find itself in the vicinity and to send it screaming to the outside.

The lifespan of a tastebud cell is about ten and a half days.

Most of the bacteria-infested water droplets carried on the wind of a sneeze travel about one-and-one-half to six feet (and about 100 feet per second) before finally coming to rest on the floor about a minute later. A small percentage, though, about 4 percent, are still found floating in the air after 40 minutes. But, because so many droplets are expelled, even the small number of bacteria-filled droplets that does not settle immediately may increase the bacterial count in small rooms by about 400–500 bacteria per cubic foot for the first few minutes. In other words, we can still catch what someone else has sneezed out some 30 minutes later, long after that person has left the room. As the medical slogan goes: "coughs and sneezes spread diseases."

Single sneezes are bad enough. When they come in sets of a half-dozen, though, they can be annoying. But when they come in sets of a hundred or more, they can be more than bothersome. They can be down right dangerous.

An 11-year-old girl admitted to the University of New Mexico Medical Center, on January 1977, sneezed this way. In fact, for three weeks, ever since she had slept with two cats, she had been sneezing at a rate of 20 times per minute. By the time the University doctors saw her, she was in great pain (her chest hurt). She was also hungry and extremely tired.

Suffering from a malady known as intractable sneezing, a rare and somewhat mysterious illness—there have been only a dozen reported cases in the last 35 years—the girl was the portrait of what happens when a reflex goes mad. The sneeze neuroanatomy is connected to a number of other regions of the brain, many of which control everything from anger to sexual pleasure. This explains why excitement or rage, or even sexual activity can trigger a sneeze. One 17-year-old girl was discovered to have sneezed for 154 days. Another girl was found to have sneezed for three years at the rate of 25 times per minute.

Mostly the onset of these bouts are brought on by psychological problems. One girl, who had been sneezing for several weeks and who had attracted media attention, stopped the minute the cameras left the room. Another began sneezing after she had moved to a new job.

As for the little girl, she was able to leave the hospital after her mother was persuaded to seek psychiatric help ("The mother was overprotective," wrote one of the doctors. The girl was "extremely submissive."). Her mother underwent therapy. And as mysteriously as it had begun, her sneezing bouts lessened. Eventually the little girl stopped sneezing altogether, some three months later.

Sneezes, like eyeblinks, have a purpose; as reflexes, they are designed to protect. Fingersnap quick, they defend the body from the incursion of foreign material. The rationale for their speed is clear-cut, obvious. But, faster than a sneeze, quicker than an eyeblink, the hiccup is a reflex, too. Yet, despite its speed, or perhaps because of it, no one knows exactly what a hiccup is supposed to do.

Each hiccup actually lasts less than a second, and would be but a brief nuisance if it only came once. But it doesn't. Hiccups, or singulti (one being a singultus), like slow drips of water from a leaking faucet, come in steady but maddening rhythmic outbursts. Usually, they pop up unwelcomed after a big meal, sneak up during a snack, or after a glass of wine. Sometimes, though, they show up for no particular reason at all. An episode can last for hours, days, even months. According to the *Guinness Book of World Records*, the longest attack lasted some 57 years.

Ruminations about hiccups have been around since recorded history. As far back as 25 A.D., Celsus of Rome wrote that a bout of hiccups was due to an infected liver. One hundred years later, the Greek physician Galen, suggested that excessive excitement provoked the stomach to violent emotions and brought on a hiccup. In his text, *Tetrabiblion*, Greek medical writer, Aetius (500 A.D.), attributed hiccups to "pungent humors" from an inflamed stomach.

During the Middle Ages, treatments aimed to cure these bothersome afflictions included everything from forcing a hiccupping patient to vomit, to offering them strange concoctions such as "rue with wine" and green ginger root soup, to stopping up a sufferer's ears with his fingers. One nineteenth-century physician, a Dr. Shortt of Edinburgh, convinced that hiccups were due to an unruly nerve, suggested the use of heated instruments to cure them. He recommended that the neck and back be burned and blistered along the nerve's course until the hiccups stopped. The record doesn't show whether anyone took up his suggestion. But, no doubt, his patients sought a second opinion.

Contemplation and mystical cures haven't yet provided a clue as to the hiccup's function. Since several vertebrates and most mammals do it, the hiccup may be a vestige of some primitive reflex, a kind of neurological appendix of the body; something our ancient ancestors had use for but which use now is lost. More closely related to the vomit reflex than to the cough, it may have been designed to dislodge food stuck on its way down to the stomach. On the other hand, an over-full stomach or a glass of alcohol can also trigger a bout of hiccups.

If science hasn't yet figured out the whys, the hows and whens are well known. A hiccup is really an involuntary and uncoordinated, almost chaotic spasmodic contraction of the diaphragm, the muscle which by its action draws air into the lungs (though other inspiratory muscles around the ribs are involved, too). When the phrenic nerve (that nerve which attaches to the diaphragm muscle) is irritated, one half of the diaphragm (rarely the entire muscle) contracts violently. An overly extended stomach triggers a burst of hiccups because the nerve that controls its movements (the gastric nerve) rubs against the phrenic nerve, essentially causing a short-circuit in the pathway. A burp triggers hiccups when the stomach pushes against or irritates the diaphragm.

The anatomy of a hiccup was uncovered in 1970 by neurologist John Newsom Davis, working at the National Hospital in London. By using sophisticated electronic equipment and some good luck, Davis measured a hiccup's duration. The hiccup, he found, goes like this:

It begins only when we are quietly inhaling, rarely when we are exhaling (the reflex actually inhibits the nerves responsible for active exhalation from firing). In fact, it is even more particular than that. It usually only pops up during the deepest part of a breath, when the lungs are most filled with air.

Irritated or stimulated, the phrenic nerve fires off a burst of nerve signals up to the brainstem and medulla. Excited, the brain fires its own signals back through the phrenic nerve. Fifty to one-hundred milliseconds later, the diaphragm muscle twitches (as do the muscles attached to the ribs, activated by their own nerves). We gasp, sucking in air. Thirty-five milliseconds into inspiration, the upper airway, or glottis, snaps shut closing off the lungs. This action produces the distinctive sound of a hiccup.

It is now some 60 milliseconds after the glottis has closed, and the diaphragm is still contracting its strongest, though the hiccup sound has long since ended. Four hundred milliseconds later, the diaphragm is still contracting, though weaker than before, fruitlessly trying to suck air into the blocked lungs. A half second after that, though, the diaphragm relaxes. The glottis opens. We exhale and relax, but only for a moment. A second or two later, after another deep breath, the cycle of events starts all over again. *Hiccup!*

The hiccup's primitive, rhythmic, stereotypical behavior can actually be used against it. Anything that can break the cycle, can usually stop a bout of hiccups from continuing. Holding one's breath often works. Sudden fright or pain produces a gasp; a sneeze or cough, which causes a sudden expiration, works too. Any one of these things may alter the cycle enough to stop hiccups from continuing.

Hiccups occur at the normal steady rate of between 5 to 25 times a minute. Sometimes, they come more frequently. People have been known to hiccup once or twice a second. When a bout of hiccups lasts longer than usual and can't be stopped, there is something wrong.

Most liver cells have a lifespan of five months.

A hiccup isn't like other reflexes. Its anatomy is spread thin. One illustration of the hiccup's bizarre nature, is that the brain can trigger a burst of hiccups by itself. Since it acts to restrain the hiccup reflex by a barrage of inhibitory signals, anything that shuts off its police-like action will start off a train of hiccups. Head trauma or brain tumors have triggered hiccups. Excessive alcohol drinking triggers them because it dampens the inhibiting action of the cerebral cortex of the brain. Without the brain to hold it, the reflex rolls on. Hiccups can also be set off when the vagus nerve, a cranial nerve that has branches that connect with the heart, throat, windpipe, intestine, and ear, is irritated (which is why granulated sugar or a tickle in the throat can turn on or shut off a bout of hiccups).

When a 27-year-old man walked into their office suffering from hiccups, Mark Wagner and J. Stephan Stapczynski, physicians at Harbor-UCLA Medical Center in California, knew that something like this was the case. The man told them that he had been hiccupping for four days straight. They came four to six times per minute and even persisted while he slept (or tried to). He was fatigued and scared.

The two physicians examined him and found nothing. The patient looked perfectly normal. Then they discovered what was wrong. Deep within his right ear, touching his eardrum and thus irritating his vagus nerve, was a tiny snip of hair. Wagner and Stapczynski washed the small hair from the patient's ear. The man's hiccups abruptly stopped.

Somehow, Descartes would have understood.

THREE

THE BODY IN SECONDS

On the afternoon of March 30 1981, close to 2:30 P.M., the President of the United States walked through the emergency entrance of the George Washington University Hospital clutching the left side of his chest. "He looked pale," one hospital aid recalled. "He looked in pain." Third year medical student, Frank Richards, remembered that the President began to stagger. And then Ronald Reagan collapsed to the floor. He had been shot.

The staff rushed to his aid, helping him onto a gurney. By now, the President was breathing rapidly, taking quick, shallow gasps. He complained of a sharp pain in his chest. His left lung had collapsed.

The hospital staff wheeled him into the emergency room's trauma area. There, doctors tore open his Navy blue business suit and the blood-soaked white shirt beneath it. Working quickly, they made a six-inch incision just underneath his left nipple with a scalpel, and carefully inserted two plastic tubes into the space between the lung and chest lining. They found internal bleeding, but his vital signs were stable. (The President's blood pressure was steady, and his pulse was fast but strong.)

Applying a weak suction in the tubing, doctors drew off the blood that had collected in the chest cavity and reinflated the President's collapsed left lung. Nurses stripped off the rest of his clothes and doctors explored his naked body for the entrance wound made by the bullet that had plunged into his chest.

Down the hall, the events that had occurred during those first 60 seconds during the attack on the President were being played out on television. Again and again, the videotape replayed the scene of the attempted assassination:

The President is waving to the cameras. Next to him, by the curb, is the black Presidential limousine, its rear door open. Then, shots.

Wounded, a Secret Service man, a policeman, and the Press Secretary fall to the sidewalk. The President is pushed into the back of the car. It speeds off. The incident lasted only a few seconds.

But the cameras had missed the other story. This one had occurred within the President's body. Indeed, that Reagan eventually survived the attack, that he did not bleed to death or die from infection, had as much to do with the biological events that occurred within the President's body in the first few seconds after the attack as it did with the skill of the doctors who later treated him. What follows, then, is that story: the first "60 critical seconds" in the life of the President.

The First Second

When John Hinkley pulls his pistol out from under his jacket, the President of the United States is only some twenty-five feet away. The gunman is standing in a crowd of reporters and cameramen. Hinkley aims his gun at Reagan and fires.

Travelling some 500 miles per hour, one of the .22-caliber bullets smashes into then ricochets off the edge of the black limousine's open rear door and plows into the President's body. The projectile enters slightly underneath the President's left armpit, then bounces off the seventh rib before finally plunging into the left lobe of the President's lung, collapsing it.

On its path, the white hot bullet tears through skin, muscle, and lung tissue, rupturing cells and shredding thousands of microscopically thin blood vessels. From the impact, blood rushes out of the torn arteries and into the President's chest lining.

From the time the round is fired to the time that it enters the President's body, only some 30 milliseconds have elapsed.

The human body is surprisingly tough. It can take blows to the torso, knocks to the head. It can be slapped, spanked, shaved, burned, paddled, and bruised. It can survive falls from ten, even twenty feet. And it can be cut. The reaction of the body to these insults, though, is always immediate.

When the President was shot, the shock to the body stimulated nerves and a host of repair systems within seconds. Clotting mechanisms rapidly slowed blood loss. Chemicals released in the area stimulated infection-fighting white blood cells. Damaged tissues activated the entire inflammation process, a complex, local defensive response by the skin and other tissues of the body to the presence of injury. Alerted, too, the brain and the endocrine glands sent forth an army of chemical messengers, hormones, into the blood to coordinate the activities of the body. Damaged, the body informed the brain of the extent and location of the destruction by a phenomenon we call pain.

Pain was triggered locally at the site of the damage when cells, shredded by the force of the impact, opened their guts to the surroundings. The sudden increase in the acidity of the area and the release of certain proteins by the damaged cells, activated an enzyme in the environment. The enzyme is called kallikrein.

Kallikrein is a fairly new member of the pain story. First discovered some 25 years ago, the enzyme caught the attention of many researchers because it was involved not only as an active player in the initiation of pain, but also in one of the most ancient diseases of man: gout.

It takes 45 minutes, from production to final secretion, for digestive enzymes from the pancreas to enter the intestine.

Back in the seventeenth century, the cane was used as much to steady an infirm body as it was to relieve the pressure from the swelling of gout. A disease of the joints, primarily of the toes, ankles, and knees, gout is a genetically-transmitted disease that primarily affects males. The pain it inflicts on the body is excruciating. "The victim goes to bed and sleeps in good health," wrote the seventeenth-century English physician, Thomas Sydenham, about the disease with which he himself was inflicted. "About two o'clock in the morning he is awakened by a severe pain in the great toe; more rarely in the heal, ankle or instep. This pain is like that of a dislocation, and yet the parts feel as if cold water were poured over them. After a time [there comes] a gnawing pain and now a pressure and tightening. So exquisite and lively meanwhile is the feeling of the part affected, that it cannot bear the weight of the bedclothes nor the jar of a person walking in the room."

The inflammation of gout is triggered when sodium urate crystals collect in the fluid in and around the joints of the legs. Urate is a breakdown product of purine molecules, one of the substances that make up the DNA molecule. Normally, urate is excreted in the urine.

But for those affected by gout, the urate crystals not only collect in the joints but, by their action, cause the secretion of a protein "factor" that, like some enticing aroma, draws every white blood cell to the area to come and feed on these crystals. The white blood cells, by the tens of thousands and the fluid with them, flow into the joints and eat the crystals thereby inflaming the area. That "factor," which attracts white blood cells, turns out to be kallikrein. As it happens, it is the action of this same molecular "factor" that attracts white blood cells into the area of damage when we are cut, bruised, or burned.

But the kallikrein enzyme, of course, plays another role, that being involved in the initiation of pain. It does this by activating another protein in the area of damage. That protein is called bradykinin.

Injured in the attack, pain nerves fire their first impulses to the brain. The signals race up the nerve bodies and head for the spinal cord. At the same time, millions of kallikrein enzymes clip off the restraining proteins attached to the bradykinin molecules nearby.

Free of their protein chains, bradykinin molecules flow to the naked tips of pain nerve endings in the region, bind to them, and further excite the pain nerves in an effort to inform the brain of the extent of the destruction.

Only some 50 milliseconds have passed since the bullet pierced his body, and the President, having not yet felt the first pain signals, is unaware that he has been shot.

More than 50 years ago, no one knew what role bradykinin played in triggering pain. Searching for the chemical components in blood, the Germans were the first to isolate the substance in the late 1930s. But they never succeeded in discovering bradykinin's role in pain. Their research got sidetracked by the Second World War.

Then, in 1949, bradykinin was discovered again, this time, though, by two scientists working in Brazil who discovered the substance purely by accident.

In his laboratory at the University of São Paulo, pharmacologist Mauricio Rocha e Silva, and his graduate student, Wilson Beraldo, were studying the venom of the Brazilian pit viper, *Bothrops jararaca,* to determine why the venom was so painful and so deadly. Rocha e Silva's first hunch was that the main pain-producing agent was histamine. Histamine is often found in the venom of poisonous snakes.

To test whether in fact histamine *was* present in this venom preparation, Rocha e Silva used a standard assay test for the chemical. He injected an anesthetized dog with the venom, then withdrew some of the dog's blood. Dripping the blood onto a fresh strip of intestinal muscle in a glass dish, the old researcher waited for the muscle to contract violently, the standard reaction to histamine. Instead, the muscle remained flaccid. Histamine was not present.

Rocha e Silva tried several more times to elicit a reaction, got nothing, and went home disappointed. When his mentor left, though, Wilson Beraldo, curious about the technique, tried it again. The intestinal muscle contracted, but not violently, the way it was supposed to do with histamine. This time, it contracted slowly.

The next morning Beraldo informed Rocha e Silva of the results and together they named the unknown chemical from a combination of two Greek words, *bradys* and *kinein,* literally meaning "slow to move." As it turned out, the jararaca venom did not contain bradykinin after all. But the two researchers were able to determine that the venom caused the slow release of bradykinin from the blood.

It takes, on average, 72 seconds for a mature egg cell to be pushed out of the ovary. The fertilized egg remains within the oviduct for about 3 days before it enters the uterus.

Since Rocha e Silva's discovery, bradykinin has been shown to crop up wherever there is pain in the body. Cuts, burns, sunburn, toothaches, headaches, sore throats, colds, allergies all involve the speedy activation of bradykinin, the chemical that not only provokes pain but initiates healing.

Still, it is in its role as the instigator of pain that bradykinin is most well-known. And bradykinin is incredibly painful. Inject a few millionths of gram (equivalent to a fragment of a single crystal of salt) under the skin, and patients will go through the roof. It is so potent, in fact, that enzymes in the blood and other tissues inactivate it almost as fast as it is formed. And, its actions are so rapid, that even the very act of inserting a hypodermic needle under the skin to take a blood sample, activates the enzymes that form bradykinin. Bradykinin triggers pain directly by causing pain nerves to fire, thereby helping to notify the brain that an injury has occurred.

The First Ten Seconds

Still active in the President's body, bradykinin binds to the walls of small blood vessels in the area. The cells making up the walls will soon pull apart leaving gaping holes between them. In roughly an hour or so, hungry white blood cells from the circulation will squeeze through these cracks and head for the damaged tissue and the alluring kallikrein there. They will feast on bacteria and the dead body parts of other cells. The fluid and the enlarged blood vessels in the area will make the skin swell and redden.

By now, the President feels a pain in his chest, though because of the excitement, he still does not realize that he has been hit.

The ancient Egyptians some 2000 years ago, didn't know about the role of bradykinin in inflammation, nor did the ancient Greeks. But the ancient scholars *did* know about the body's reaction to injury. Even Hippocrates, around 400 B.C., knew that before there could be healing and repair, there must be inflammation.

Inflammation was first described in detail by Aulus Cornelius Celsus, a Roman science-writer of sorts (perhaps the first) who, around 30 A.D., collected Greek knowledge for the reading pleasure of Roman audiences. In one of his medical texts he described the classic signs that characterized inflammation: *rubor* (redness), *calor* (heat), *dolor* (pain), and *tumor* (swelling). One hundred years later, the Greek physician, Galen, who had served as a surgeon at the Roman gladiatorial arena and probably saw his fill of killing and injury, added yet another signature to inflammation's roster, *functio laesa*, altered function.

But, though the ancient Greeks and Romans associated the signs of inflammation with injury and infection, they wrongly regarded the process as simply part of a painful illness. It wasn't really until the advent of the microscope, and studies on living tissues in the late nineteenth and early twentieth centuries, that scientists truly began to understand that inflammation was the first, early stage in the body's reaction to injury.

Rubor, calor, dolor, and *tumor;* the response in the body to injury is universal. From frigid cold, searing heat, damaging X-ray radiation, to bacterial invasion; the same events occur during the inflammation response whether the damage occurs within the body or on its surface.

Just to demonstrate the inflammation reaction's speed and ubiquitous nature, run a fingernail quickly but firmly down the length of the skin of your forearm and watch what happens. Immediately, a white line will first appear. The reaction is due to the action of the blood vessels in the area which squeeze tight in the mistaken anticipation that blood will be lost.

After some 7 to 10 seconds, a red streak will appear *(rubor)*, the result of the dilation of these same blood vessels allowing more warm, red blood *(calor)* to pass. And there will be some pain *(dolor)*. Seconds later, you should see a distinct ridge *(tumor)* along the red streak where the vessels have leaked fluid into the tissue spaces.

A breath lasts about five seconds; two seconds of inhalation, three seconds of exhalation.

In the 1940s, English physician Thomas Lewis, believed that the reddening and the swelling that occurred in this reaction were due to the release of something he called "H substance" from the blood, and he proved this by a very simple experiment.

Lewis raised his arm to the ceiling and waited for the blood to drain out of it. Just before he began to feel his arm becoming numb from the loss of blood, he tied a tourniquet around his upper arm to prevent any more blood from reaching it. Then, he scraped his near bloodless arm with his fingernail.

Nothing happened. The scratch simply remained there. There was no redness, and no raised weal. Surprised, he lowered his arm and released the tourniquet. Sure enough, when the blood flowed back into his arm, he saw a large red ridge appear where he had scratched his skin.

Lewis concluded correctly that the inflammatory "H substance" must have been released when he scratched himself, but, in order for it to show its effects, blood had to flow back into his arm. Since the red weal appeared immediately after the tourniquet was released, he reasoned, the "H substance" must have prepared the blood vessels already and only an inrush of blood was needed to show up the effect. As it turned out later, Lewis's "H substance" contained a large portion of bradykinin but mostly a chemical identified as histamine, the chemical mediator of the allergic reaction.

Lewis's mysterious "H substance," histamine, might never have been fully understood or localized in the human body were it not for the discoveries made by the British physiologist Henry Dale and the German bacteriologist Paul Ehrlich. In 1877, Ehrlich was looking under his microscope at some bits of skin when he spotted a curious looking cell. It appeared to be stuffed with large, dense granules. And when he soaked them in special dyes, they stained brilliantly. He assumed, incorrectly, that this giant cell had engulfed the granules in some feeding frenzy, and so called the cells, "mast cells," from the German *mast* meaning "a fattening feed." The name stuck, even though the granules are, in fact, not engulfed but rather made by the cells. The granules—a thousand or more of them are packed into each mast cell—are filled with the chemical histamine.

When Ehrlich first made his discovery of the mast cell, Henry Hallett Dale was only two years old. Born in London on June 9, 1875, Dale eventually would receive the Nobel prize for his work on the chemical transmission of nerve impulses. But during his late twenties, he was only a fledgling pharmacologist employed by the Wellcome Physiological Research laboratories in England. There, the future Nobel prize winner studied the chemical composition and effects of ergot, a natural product derived from the fungus *Claviceps purpurea* that grows wild on rye grain.

Then, in 1910, Dale isolated a compound from ergot that

triggered a powerful allergic reaction in his laboratory animals. When he injected a tiny bit of this chemical into his laboratory animals, they nearly died of suffocation. The compound triggered the muscles of the animals' air passageways to squeeze tight, shutting off air to the lungs. The material Dale identified was histamine. Years later, scientists discovered histamine in Ehrlich's mast cells and in several other cells of the body.

As Dale found out, histamine is a powerful substance. Not only does it rapidly trigger the smooth muscles of the intestines and the air passageways to contract violently, but it also prompts blood vessels to distend and induces their cells to separate. Blood fluids flow into the tissues. Vessels swell. The result: our eyes get red, our nose gets stuffy. And, sometimes, our only recourse against this allergic reaction is to take an antihistamine to block the actions of this chemical and to clear up its symptoms.

Moments after bradykinin is released from the site of the President's wound, it diffuses to the mast cells in the region and binds to them. Those mast cells already in the area respond immediately by dumping out large quantities of histamine. Other mast cells will soon arrive and they too will spray their histamine upon the damage.

Molecules of histamine and bradykinin bind to the blood vessels in the region. The vessels split their seams further allowing more fluid and even more molecules of kallikrein and bradykinin to pour into the area.

When it is released by injury—mechanical, thermal, bacterial, or chemical—bradykinin not only sets off part of the pain response, it also triggers mast cells to release their histamine. The response is immediate, or nearly so. The small blood vessels near the injury become leaky so that the area is bathed in fluid and infection-fighting blood cells. The surrounding tissue swells. By doing this, histamine and bradykinin set the stage for the initial phase of healing and repair.

In this battleground of injury, though, histamine and bradykinin are not the only warriors. There are many, many others. But two, in particular, play a major role in the inflammation response, and they were both discovered by the same researcher.

In 1931, the German researcher, U. S. von Euler, isolated a small protein substance from the brains of his laboratory animals that he concluded was part of the secretions of the nerve cells. He called this crude material, substance P, for the *p*reparation that contained this active extract.

Then four years later, von Euler found a different set of chemical substances in the fresh semen of animals. He mistakenly thought these chemicals were made in the prostate gland of the male reproductive tract. The name he gave these chemicals, prostaglandins, has been used ever since his discovery.

Substance P and the prostaglandins come into play during the first few seconds after injury. When bradykinin triggers the pain nerves to fire, the pain signal travels to the spinal cord where substance P is released like a neurotransmitter, the chemical that carries signals between nerve cells. Other nerves at the spinal cord, then, send the pain signal the rest of the way to the brain. But substance P may also be released by the nerve terminal endings at the site of damage. There substance P may coax mast cells to release even more histamine into the region.

While molecules of bradykinin are busy exciting nerve cells, they also trigger the release of the prostaglandins from their home in the membranes of the cells of the blood vessels in the region. The prostaglandins, in turn, like the molecules histamine and bradykinin, promote tissue swelling. The area stays red and tender until infection is suppressed and healing starts.

Fingerprints form six to eight weeks before birth.

The First Fifteen Seconds

In the back seat of the Presidential limousine, Reagan feels the pain of the wound. He is bleeding. The left side of his white cotton shirt is wet with blood.

When the skin is pierced, it bleeds. When a major blood vessel is cut, it bleeds profusely. To stay alive, though, the bleeding has to be stopped. The normal workings of the heart and the rest of the circulation depend on it. Severed arteries or crushed blood vessels allow bleeding, and bleeding lowers blood pressure, or fluid volume in the circulatory system. Too much blood loss over a short amount of time and the system goes into shock. Vital organs soon begin to fail for lack of nourishment and oxygen.

The brain fails first. Without an adequate supply of oxygen and glucose sugar, the brain shuts down. An individual would first feel faint. Then, he would lose consciousness in about 30 seconds or so.

The kidneys would be the next to go. Because they rely on the normal blood pressure of the circulation to do their job, a drop in blood pressure would shut them down. That is why it is essential to stop the bleeding at the earliest possible moment.

The process by which blood loss is stemmed is called hemostasis. Whenever a vessel is severed or ruptured, a series of events occur that, in effect, seal off the damage and stop the flow of blood out of the vessel.

Within two seconds after a blood vessel is cut, the wall of the vessel reflexively contracts in a vascular spasm. Arteries, in particular, which have a thick muscular wall, can automatically shut down and limit the outflow. The more the vessel is traumatized, the greater the spasm. The spasm may last as long as 20 to 30 minutes, during which time the ensuing processes of hemostasis take over.

During this time, specialized cell fragments called platelets arrive at the scene. Platelets are the smallest formed cellular elements of the blood. The red cells carry oxygen, the white blood cells defend the body against bacteria and viruses. The platelets play a central role in hemostasis. They clump together to form a plug that temporarily seals any break in a blood vessel.

Every drop of blood contains about a quarter of a billion platelets. Since the average human body harbors some four to five quarts of blood, it figures that there are about a trillion or more platelets in the blood at any one time.

But platelets are not cells; they are fragments of cells called megakaryocytes. These are huge platelet-mother cells found within the bone marrow. As each of these platelet-mother cells matures, its cytoplasm (the material outside of the cell's nucleus) breaks up to form several thousand platelets. The resulting platelets, roughly a third the diameter of red blood cells, come out looking a bit like droopy Frisbees.

Platelets have no nuclei and no DNA. That makes it nearly impossible for them to manufacture proteins to maintain their integrity. Thrown out of the bone marrow, they live for only about 10 days, less if they happen to be consumed in the line of duty. Although they don't contain a nucleus, they do contain various types of sacs or granules filled with an array of important chemicals to do their job. Some of the granules contain the chemical called adenosine diphosphate, or ADP. Others contain various protein factors which play a role in the clotting reaction.

Normally, platelets flow along the body's blood vessels like bits of debris. Occasionally, they rub alongside the vessels' smooth lining in their journey through the body. But, this does not activate them. Only when a vessel is cut and blood flows out through the wall of the vessel, do the platelets wake up.

The stimulus that activates these cell fragments lies within the matrix of the blood vessel wall. This material, called collagen, is the same substance that makes up the soft part of the nose and the ear. But, it is also the material that is first exposed to the platelets when a vessel is damaged.

Activated, platelets in the flowing blood adhere quickly to the exposed collagen, and then other passing platelets adhere to them. Within ten to fifteen seconds, if the trauma to the wall is severe, and in one or two minutes if the trauma is minor, a plug of platelets builds up.

The wall of the cut blood vessel also contains an enzyme called "tissue factor," which initiates blood clotting. Within the fluid of the blood there are a number of other protein factors or enzymes. Exposed to "tissue factor," they become activated, but in a highly ordered way. No sooner does "tissue factor" activate one of these chemical factors, then that excited protein factor goes off to activate another. Like falling dominos, eventually, the last plasma enzyme to be activated changes an enzyme called prothrombin into its active form called thrombin. Thrombin, in its turn, converts the plasma protein fibrinogen into strands of fibrin, which then form a webbing that traps platelets and blood cells. Thrombin also makes platelets sticky and they stick to the cut. While all of this is going on, the platelets themselves release their store of ADP, which causes more of these cell fragments to become even stickier and activates other platelets to release their ADP so that even more platelets join the fray.

In hemostasis, platelet secretion and platelet aggregation work together. When the individual platelets adhere to the collagen in the cut blood vessel wall, they secrete their store of ADP. The ADP then stimulates many more of these clotting fragments to swarm into a lump at the cut vessel. Passing platelets adhere to those already bound to the collagen. Squeezed out by the platelets, as well, the chemical, serotonin, triggers the local blood vessels to narrow, thereby helping to restrict the flow of blood. Thus in hemostasis, a number of substances in the blood vessel wall, in the blood plasma, and in and on the platelets themselves, get into the act. The ultimate result: Platelets form a solid plug that arrests bleeding.

Men have four to five erections a night (when they are asleep), each lasting from five to ten minutes.

Normally, all of these factors are present in the body and so clotting takes place. But, in some people, the clotting mechanism is impaired. Their body either does not make a necessary "factor," or the "factors" they have are damaged in some way. Those who are in this sorry state run the risk of harmful if not lethal bleeding unless they avoid the normal hazards of life. The hereditary disease known as hemophilia is the most famous example of such a "bleeding disease."

The defect in hemophilia is in the clotting pathway. A protein called antihemophilic factor is missing or has a markedly reduced activity. Antihemophilic factor is essential for the main pathway that converts prothrombin to thrombin. Without this factor, clotting occurs very, very slowly, sometimes over a period of days.

This strange affliction is transmitted by mothers to their male children through a sex-linked recessive gene. Thus, women almost never get the symptoms; it usually strikes only males, although it does not necessarily strike all of the male members in a family.

Hemophilia is an ancient disease, but one that was well-known to the ancients. Indeed, the Egyptian Pharaohs decreed that it was forbidden for a woman to bear children if her first-born son bled to death from a scrape or small cut. The ancient Talmud barred families from performing circumcision if two successive male children bled to death from the procedure.

The most famous carrier of the disease was Queen Victoria, who unknowingly transmitted it to the royal families of Prussia, Spain, and Russia. The youngest of her four sons, Prince Leopold, Duke of Albany, had hemophilia. And two of her five daughters, Alice and Beatrice, were hemophilic carriers. Alice transmitted the disease to her daughter, Alexandra. Alexandra later married Nicholas, the last Tsar of Russia.

Alexandra was twelve years old when her uncle, Leopold, fell, suffered a minor blow on the head, and later died, at thirty-one, of a brain hemorrhage. This might have served as a warning to her not to have children, since she was likely a carrier of the disease, but, it didn't. Partly because she little understood the hereditary pattern of the disease, and partly because hemophilia, like pneumonia and smallpox was simply considered one of the hazards that royal families had to face, she bore children for her husband.

Thus, on August 12, 1904, Tsar Nicholas II wrote in his diary: "A great never-to-be-forgotten day when the mercy of God has visited us so clearly. [Alexandra] gave birth to a son at one o'clock. The child has been called Alexis."

But, six weeks later, the Tsar's mood changed from happiness to despair. "A hemorrhage began this morning without the slightest cause from the navel of our small Alexis," wrote the Tsar in his diary. "It lasted with but a few interruptions until evening." In the days that followed, young Alexis began to crawl and toddle in his crib. When he fell, small bumps and bruises appeared on his arms and legs. A few hours later, the bruises grew to massive dark blue swellings. Upon seeing this, the Tsar and his wife could not help but fear the worst. Their only son, the heir to the Russian throne, had hemophilia.

Alexis died on July 4, 1918 at the age of 14, but not from hemophilia. A victim of the Russian revolution, Alexis, along with his four sisters, the Tsar and his wife, and their servants, was shot to death. His body, along with the bodies of his family, was burned and later thrown down a mine shaft. In life, poor Alexis suffered for the lack of a single clotting "factor." His body could not stop from bleeding.

By the same token, when clotting occurs all too frequently, that can be bad, too. The clotting process must occur rapidly at the moment the injury to the system occurs. But it must also be precisely regulated. There is a fine line between hemorrhage, uncontrollable bleeding, and thrombosis, uncontrollable clotting.

If a clot occurs spontaneously, the person runs the risk of plugging his or her own circulation from the inside. Because platelets can stick to one another and to the wall of a blood vessel, and because they promote clotting, they can form clots or thrombi and impede the flow of blood. Thrombi often form in veins, particularly in elderly bedridden people who have undergone surgery. They are dangerous primarily for their tendency to break loose becoming emboli, or detached clots, that can be swept by the bloodstream through the heart. If they lodge in a small artery feeding the heart, they cut off blood flow and lead to myocardial infarction, the death of part of the heart muscle; a heart attack. If they lodge in the brain, reducing the blood supply there, they can trigger a stroke.

But, for most of us, clots form only when we need them and not when we don't. The complex system of the "factors" themselves are the reason for this. Like the soldiers in charge of launching a nuclear missile, one "factor" cannot set off the system without the aid of another. Both cannot launch the missile without the aid of yet another, and so on. Moreover, the clotting factors have a short lifetime because they are diluted by blood flow, removed by the liver, degraded by enzymes, and inactivated by specific inhibitors. This whole complex system assures that a clot will only form where and when it is supposed to.

The most important inhibitor of the clotting mechanism is antithrombin III, which binds to and inactivates a number of the clotting "factors." The inhibitory action of antithrombin III is enhanced by heparin, found in mast cells near the walls of blood vessels. Heparin acts as an anticoagulant by increasing the rate of formation of the complex between thrombin and antithrombin III.

An exposure of about a tenth of a second is required to establish a good image on the retina. Moving pictures are shown at a rate of about 24 frames per second, too fast for humans to see individual images. Images fade one into another and we get the illusion of motion on the screen.

Using clots to stop a small nick in a blood vessel works well in most cases. But, unfortunately, the body cannot stop a major hemorrhage. Blood loss is so rapid that a clot dislodges as fast as it forms. Instead of relying only on the clotting mechanism, then, the body adjusts to the loss of blood by reallocating the blood that is still left in the body to other, more important regions. Thus, when the blood pressure in the body begins to drop, blood is shunted to organs like the heart, liver, and lungs, and away from organs like the intestine and the skin. This is why patients with hemorrhage go white and feel cold to the touch.

The First Sixty Seconds

The limousine heads for the White House, but Reagan, feeling the pain in his chest and the blood on his shirt, directs the driver to head for the hospital.

He feels slightly faint and a little dizzy. He feels his heart. It is beating fast.

During the latter part of the nineteenth century, French physiologist Claude Bernard developed the concept that all life regulates its own internal processes. Perhaps American physiologist Walter Cannon took note of Bernard's philosophy. At the turn of the century, Cannon was interested in the actions of the nervous system on internal secretions. The American physiologist directed his attention to a response which he called the "emergency reaction."

When they confront a threatening or dangerous situation, animals have only one choice: they can either put up a fight, if the opponent is weaker or evenly matched, or they can flee, if the opponent is larger or faster. Either way, this so-called "fight-flight" or emergency reaction is reinforced by physiological changes within the body. Physical trauma or mental agitation triggers the secretion of hormones from the adrenal glands and the pituitary and prepares the body for action.

But, the nervous system is activated, as well. As little as 50 milliseconds after a trauma, nervous signals from the brain travel down the spinal cord and activate the glands, organs, and muscles of the body. They do this by squirting out the chemical transmitter norepinephrine.

As a neurotransmitter involved in the emergency reaction, norepinephrine has a wide range of actions—all of which help the body survive the stressful situation. By locking onto chemical receptors of the heart, for example, norepinephrine increases the rate and strength of heart contraction. The neurotransmitter also makes the blood vessels constrict, thereby raising blood pressure. Activated by this chemical, the liver

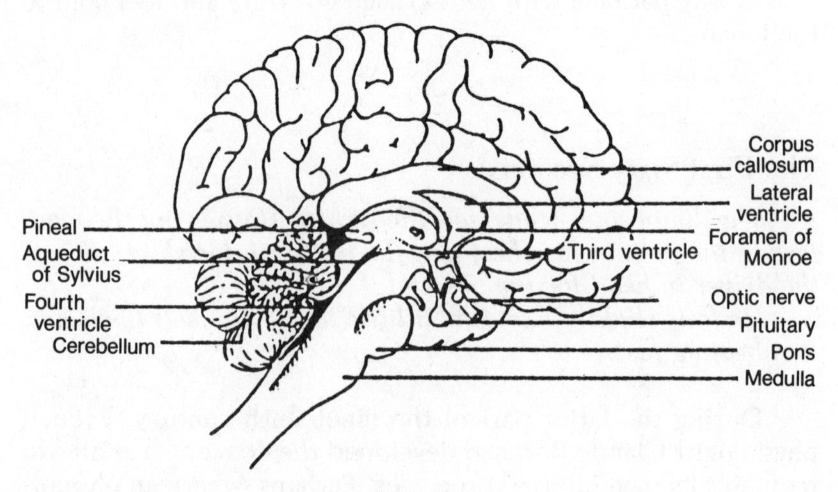

The brain.

breaks down its store of sugar and fat and releases them into the blood for fuel to stoke the fires of metabolism. Because of norepinephrine's action, blood is shunted to the muscles of the arms and the legs and away from the intestines. Digestion slows, but, the heart beats faster. The arms feel stronger. Yet, the action of norepinephrine released by the nervous system is short-lived; its effects last less than a second.

The Seventh Cavalry in all of this is the system of stress hormones released by the body's endocrine glands. In a sense, the secretions of the nervous system act like the first wave of shock troops at the start of battle initially activating the body, while the hormones, which arrive later, behave like the reinforcements. They carry on the battle where the nervous secretions have left off.

For example, the adrenal medulla, the endocrine gland resting over the kidneys, secretes adrenalin and also norepinephrine. These two chemicals, though, flow out into the blood circulation slowly. Whisked away by the blood they then take some 20 to 30 seconds to reach the heart, the muscles, the liver, and the brain. But their effects on the body last seconds, sometimes minutes longer than do the effects of the substances released by the nervous system. The hormones are removed from the circulation slowly.

The pituitary gland and the bit of brain tissue above it, called the hypothalamus, also release their store of hormones. The hypothalamus dumps a chemical hormone into the blood called vasopressin. Thirty seconds or so later, vasopressin is at the kidney. Its role there is to coax the kidney to dump water back into the circulation rather than to dump it out as urine. In doing so, vasopressin's actions on the kidney raises blood pressure. Were the body ever to be cut during battle, the increased blood pressure would be vital to the survival of the body. Even with a small loss of blood, the organs of the body could be maintained.

At the same time, the pituitary gland dribbles out its store of a chemical known as adrenocorticotropic hormone, or ACTH. Twenty or thirty seconds after it is dumped into the blood circulation, it arrives at the adrenal gland. There it stimulates the adrenal cortex (the part that drapes over the adrenal medulla). The adrenal gland, in turn, secretes its own store of

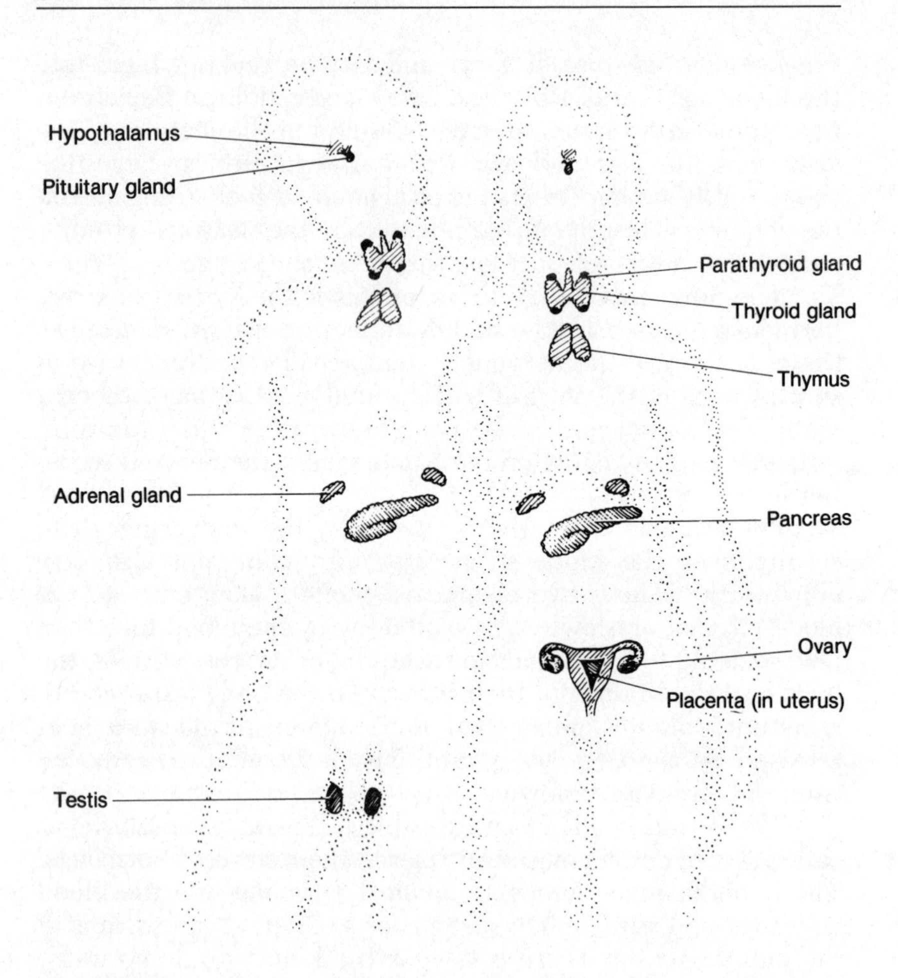

The positions of a few of the hormone-secreting endocrine glands of the body.

hormones which, after a time, travel to every part of the body and help to repair the damage done.

But it would be a sadly inefficient system, indeed, if only the body, and not the brain, were to receive the beneficial effects of these hormones in times of stress. The brain needs to be kept informed of the activity elsewhere in the body. As it happens, the brain is visited by both vasopressin and ACTH. But their actions on the brain are far more dramatic and long-lasting than one might expect.

Back in the late 1950s, David de Wied and his colleague Peter Smelik, working at the University of Groningen in The Netherlands, were studying rats to prove that the release of vasopressin from the hypothalamus prompted the release of ACTH from the pituitary gland below it. In an attempt to show this, they cut out that part of the hypothalamus that secretes vasopressin. Then, they exposed the rats to a varied number of brief, but stressful, situations.

The results surprised them. When the animals were exposed to physical trauma, such as an injection of histamine, the rats secreted ACTH in normal amounts. But, when they were put into a strange cage or exposed to loud noises—all *emotionally* stressful situations—ACTH secretion dramatically decreased. Somehow, these vasopressin-deficient animals were not responding to emotional stress the way normal animals did.

To test their observation, David de Wied placed his rats into a specially designed cage called a "shuttle box." In this box, rats learn to jump over to the other side of the cage when they hear a bell in order to avoid a mild electric shock coming from the metal grid on the floor. After they learn this, they have to remember to jump even when the electricity is turned off and the bell rings.

When de Wied removed the vasopressin-secreting part of the hypothalamus from his animals and then tested them in the shuttle box, the animals learned to avoid the foot shock well enough, but they didn't retain this information. They forgot quickly. But, when de Wied injected vasopressin into these animals and tested them again, they behaved just like normal rats and remembered to jump long after the electricity was turned off.

David de Wied and other scientists have since found that both vasopressin and ACTH enhance memory in people as well as in rats. In fact, scientists have injected a synthetic chemical called DGLVP, a substance that mimics vasopressin but that does not activate the kidney, and an analog of ACTH, called ACTH 4-9, that does not activate the adrenal gland, into rats and have shown that these chemicals, too, increase memory retention.

Apart from their obvious clinical value—they have been used to improve the memory of the elderly—they have an important evolutionary value, as well. Through the millennia,

ACTH and vasopressin have evolved as messengers, not only for the body, but for the mind. Both are released immediately during stress, mental or physical. And we generally remember these stressful or embarrassing situations long after we forget the more mundane events in our lives. Researchers like de Wied think that the "higher function" for these hormones evolved to keep the brain informed when something happens that is important for survival. So we remember. And we learn.

All these processes—the clotting of blood, the initial process of inflammation, the pain, the secretion of hormones, the laying down of memory—occur in the first 60 seconds after injury. All of these processes, the products of a thousand million years of evolution, saved the President's life that March day in 1981. But, then, all these save the lives of all of us, every day.

The black Presidential limousine speeds off to the hospital. Still bleeding, Reagan clutches the left side of his chest. It is painful and very sore. Leaning back against the car's seat cushion, the President looks out of the smoke-colored window, the terrifying image of the man with the gun lodged in his mind.

FOUR

THE BODY IN MINUTES

Standing half-naked in the drab room of the British Interrogation House near Luneburg, Germany, Heinrich Himmler, the infamous Reichsfuhrer of the Nazi SS, looked worn and tired. By taking on a false identity, he had eluded capture by the Allies for days. But now arrested by them, he was being treated like a common prisoner.

On this the 23rd day of May 1945, British Intelligence officers were about to question him at length, but those interviews never took place. Frightened when a doctor ordered him to open his mouth for inspection and spotted "a small black knob sticking out between a gap in the teeth," Himmler suddenly turned his head aside and bit down on the small glass phial of cyanide in his mouth. He died fifteen minutes later.

Had his jailers been faster with the stomach pump and other antidotes for this poison, they might have had time to save Himmler from death. Cyanide's first symptoms, giddiness, headache, and palpitations, don't appear for some ten minutes after the poison is swallowed. In the world of drugs and poisons, though, cyanide's lag time is certainly no anomaly. A tablet of aspirin may silence pain after twenty minutes; a capsule of Valium may destroy anxiety after sixty. A whiff of laughing gas may numb the brain after only one.

Usurpers and criminals alike have known about the timely actions of drugs for centuries. Even the word, assassin, comes from the ancient Arabic, meaning "one drugged on hashish." Many of the poisonings of ancient, medieval, and even modern times were done by placing a small measure of the slowly acting arsenic into the food and drink of a rival or contender for the throne. A single fatal dose of the material usually caused death within one to twenty-four hours. The subtle and cunning Livia used a slow-acting poison to kill her husband, the Roman Emperor Augustus. He died later that day. But, a cup of poison Hemlock was administered to the great Greek philosopher Socrates to kill him quickly. Death occurred in a matter of minutes. Later, death by Hemlock was decreed for Aristotle. And the poison would have been used on him if Aristotle hadn't quietly escaped the city of Athens.

Eyelashes, which are more plentiful on the upper eyelid, are shed continuously. Each of the more than 200 hairs per eye lasts from three to five months.

In her palace in Alexandria, the Egyptian Empress Cleopatra used her physician, Olympus, to prepare a pharmacy of poisons, then watched as the poison-maker tested the concoctions on her prisoners. From her "experiments," she found that some drugs were quite slow to act but caused great pain. One, she found, was particularly rapid. It killed within minutes. Unfortunately, however, it caused convulsions and left the faces of its victims distorted after death. From what toxicologists know now, the drug was probably strychnine, a chemical found in the seeds of the *Strychnos Nux Vomica* tree that grows in India. The fast-acting poison's first symptoms before a horrible death—convulsions, body spasms, and the facial distortion known as *risus sardonicus*—appear some ten minutes after ingestion. Perhaps because of the rapid, though painful death from *nux vomica* (strychnine), Cleopatra chose to take her own life by allowing a poisonous asp to bite her. Her death was rapid, but painless.

Cleopatra may have been one of the first to realize that a drug taken orally is slower to act than one that is injected through the skin. The relatively slow onset of action from an orally administered drug is partially due to the fact that the medicine must first travel into the stomach and then usually pass through the lining of the small intestine before it enters into the circulation. That takes time. A drug injected under the skin, as Cleopatra's snake demonstrated, has a much faster action simply because it avoids the barriers of the gut wall.

Still, several orally administered drugs are able to make the journey past the gut more quickly and easily than others. Alcohol, for one, avoids the intestine altogether. On an empty stomach, it is absorbed directly through the stomach lining and travels right into the bloodstream in less than a minute.

Alcohol's speedy rate of penetration through the lining of the stomach is determined by its rate of diffusion. The phenomenon can be demonstrated this way: A drop of blue dye into a glass of water will, after a time, diffuse throughout the water. The liquid will appear to be colored a light blue. The dye-stained water is due to the frantic vibratory movements of the dye particles as they travel from where they are most concentrated to where they are least concentrated. Indeed, the more concentrated the dye solution, the more the particles will bump into each other and the faster the dye particles will move out into the surroundings.

Thus, the rate at which alcohol is absorbed into the bloodstream depends on its concentration. The more concentrated alcohol is in the stomach, the more rapidly it is absorbed into the bloodstream and the faster its action. A beer, which only contains some 12 percent alcohol, is absorbed more slowly into the bloodstream than, say, a shot of whiskey. Whiskey contains about 40 percent alcohol. It gets into the bloodstream that much faster.

Separated by a membrane it is able to penetrate, alcohol will still distribute itself in this fashion. Alcohol is a small molecule that is soluble in both water and fat and thus can travel through membranes and cells without difficulty.

By the same token, because alcohol is so readily diffusible through membranes, once in the bloodstream, some of it is excreted out of the body through the membranes of the lungs. A drunk's "alcohol breath" is the result of this exhalation of alcohol out of the lungs. The alcohol has simply moved from a region of high concentration (the blood) to a region of lower concentration (the air of the lungs).

But alcohol is atypical. Not only does it dissolve in both fat and water, it is a liquid and thus is already in solution. Most orally administered drugs are solids (to retain their effectiveness) and therefore have to first dissolve in the stomach before they are absorbed into the bloodstream. Thus is the fate of aspirin.

The severed finger tips and nails (above the first crease of the first joint) of children below the age of 12, can regenerate, and can do so in about 11 weeks. Adults do not have this ability.

Taken by mouth, the rate of aspirin's absorption through the gut depends on the speed at which the pill itself crumbles in the stomach. But aspirin's rate of absorption (and many other drugs) is also dependent on how fast the stomach empties. In solution, though, and in an empty stomach, its absorption rate is rapid. It starts to be absorbed into the bloodstream in as little as a minute or so, though half of it is still in the stomach fluids twenty minutes after ingestion.

The nature of the stomach is such that it allows weak acid molecules, like aspirin, to pass directly through the stomach and into the circulatory system. However, drug molecules such as codeine and morphine, which are weak bases (or alkaline in nature) cannot pass through the stomach wall. They must survive several hours of exposure to the harsh acid juices of the stomach before they are passed into the intestine where they may be absorbed into the bloodstream slowly. This is the main reason why drugs that are weak bases, like morphine and heroin, are usually injected directly under the skin and into the circulation.

In the spring of 1943, it was not Albert Hoffman's intention to swallow some of his lysergic acid diethylamide. It was an accident. Weakly acidic, the substance readily passed through the Swiss chemist's gut and into his body. Lysergic acid diethylamide is more commonly known as the psychedelic drug, LSD. Hoffman had first extracted the molecule from ergot, the chemical from the fungus parasite that grows on rye grain, the same fungus that Henry Dale had found earlier to contain the chemical histamine.

But unlike Dale, Hoffman, working at the Sandoz Laboratories in Basel, Switzerland, had synthesized the LSD chemical in an effort to determine if it could be used for medicinal purposes. He was well aware that when people ingested the ergot-contaminated rye grain, they often suffered from bloodless gangrene of the limbs brought on by the substance's ability to constrict blood flow to the arms and legs. Victims also had the terrifying feeling that their skin was burning. During the Middle Ages, the epidemic was called St. Anthony's Fire, a reference to the shrine of St. Anthony where relief from the blistering pain was sought.

His intention, though, was not to scare people but to use the LSD for the treatment of migraine headache and to control the bleeding of women after they gave birth. But, oddly, when he tried out the chemical on his laboratory animals, the LSD had little if any effect on them. So the drug was put on a shelf and forgotten.

Then, on the sixteenth of April, 1943, Hoffman accidentally consumed some of his LSD and was gradually "seized by a peculiar sensation of vertigo and restlessness." He drove home, drew the curtains, fell into bed and began to hallucinate. Two hours later his trip was over.

Frightening though the experience might have been, Hoffman later decided to try the drug again. Taking a dose of the stuff into his mouth, he waited for the effects to take hold of his mind. For thirty minutes he sat in his lab, feeling fine, then: "after 40 minutes, I noted the following symptoms in my laboratory journal: slight giddiness, restlessness, difficulty in concentration, visual disturbances, laughing. Later . . . space and time became more and more disorganized and I was overcome by a fear that I was going out of my mind." His hallucinogenic journey lasted about 10 or 12 hours.

If Hoffman had a spectacular voyage, so too did the LSD molecules he had swallowed. All drugs, after they leave the confines of the gut, are swept up by the bloodstream which distributes them evenly throughout the circulatory system in about sixty seconds. The speed of dispersion is due to the action of the heart, which pumps about five quarts of blood each minute.

Dispersed throughout the body, a lot of the drug ends up in places where it was not intended to do its work. For example, once cocaine is sniffed up into the nose and is absorbed directly into the bloodstream by way of the delicate membrane lining the nostrils, the "high" lasts only a few moments. The reason: Most of the drug is broken down by the liver in about five to fifteen minutes.

Once they are in the body, drugs can wind up almost anywhere. Nicotine is often found in breast milk. The active ingredient of marijuana, called THC, can be found in the liver, but also in the kidneys, spleen, lungs, and even in the testes. Moreover, depending on its chemical makeup, a drug could be excreted in everything from urine, feces, sweat, semen, and saliva, to an exhaled breath.

For a drug to have its desired effect, though, it has to travel past the confines of the capillaries, the microscopically thin blood vessels of the body, and into the tissues, for it is only at the capillaries that drugs are exchanged between the circulation and the rest of the cells of the body. For most drugs, that is not really a problem. The cylindrical walls of the capillaries are only one-cell thick and the cells are not tightly glued together. There are pores or spaces between them. It is here, through these spaces, that drugs travel into the tissues.

Drug molecules are smaller than the pores they have to travel through. As a result, drugs are able to pass through the capillaries and into the tissues rather easily and quickly. The speed at which they do so depends (as it did with alcohol through the stomach) on their concentration in the blood. The more concentrated a drug is in the bloodstream, the faster it will diffuse into the tissues of the body.

Some nerves transmit signals for pain at just one-half mile per hour. At that rate, an impulse would arrive at the spinal cord a second or so after stubbing a toe.

Guided by such evidence about the easy movement of drugs through the walls of capillaries, physicians and chemists of the nineteenth century believed that it was by this route that all medicines travelled into the brain. How better to explain the psychedelic actions of drugs found in certain mushrooms than to say that their effects on the mind were the result of the drugs slipping into the brain through pores in the capillaries? But the ideas of the past gradually disappeared as the true nature of the brain's environment became clear.

Some twenty years after he had discovered the mast cell in 1877, German biologist, Paul Ehrlich, developed the idea that the brain might be surrounded by a barricade of sorts that keeps cells and most materials in the blood out of the brain but selectively lets only some material through. He proposed this concept of a blood-brain barrier (a BBB for short), after a rather simple experiment. He injected harmless colored dyes into the bloodstream of his laboratory animals and found that the dye stained every organ and tissue of the body save one, the brain. The dye molecules were small enough to get through most capillary pores but were not able to pass through the capillaries in the brain.

Ehrlich's discovery, and the later discoveries to follow, were instrumental in determining just how the brain selectively barricades material from entering it. First, the capillaries in the brain are not like the capillaries elsewhere.

Instead of possessing holes and pores, the cells that make up the brain capillaries are sealed together tightly. There is not so much as a crack between them. Furthermore, the capillaries are encased by a thick fatty sheath made from the extensions of nearby star-shaped astrocyte cells.

Shut up tight, the brain can be selective in what it allows to pass into it. For example, some drugs, penicillin, for example, and many components in the blood have a nearly impossible time getting into the brain even though they are smaller in size than the average capillary pore. To enter the brain, many drugs and components of the blood have to be carried in by highly selective transport systems positioned at the entrance to the barrier. This is how sugars and some amino acids get into the brain. And it is also why, for many years, patients suffering from Parkinson's disease were so difficult to treat.

When, in 1817, James Parkinson, the English physician working in London, first described the motor disorder that now bears his name, he was unaware of its true nature. The cause of the involuntary tremors of the limbs and head that his patients suffered was a mystery. The enigma was cleared up more than one hundred years later when, in 1966, Oleh Hornykiewicz, examined the postmortem brains of patients who had had Parkinson's disease. He discovered that these brains had low levels of the neurotransmitter dopamine. The treatment was obvious. Inject patients with the dopamine they lacked and the symptoms will clear up.

The problem was that it didn't work. Dopamine doesn't cross the blood-brain barrier and so doesn't get into the brain where it is needed. The solution, Hornykiewicz reasoned, was to find something that did. His answer was to inject into his patients the chemical from which dopamine is made, L-DOPA. Unlike dopamine, L-DOPA readily crosses the blood-brain barrier and is converted into dopamine in the brain. The symptoms of his Parkinson's disease patients cleared up and L-DOPA has been used ever since as a treatment for some individuals suffering from this disease.

A barricade sealed tight against the onslaught of foreign chemicals is a brilliant evolutionary concept. The brain and the fluid it is bathed in, called cerebro-spinal fluid or CSF, can be protected and carefully cushioned from unnecessary

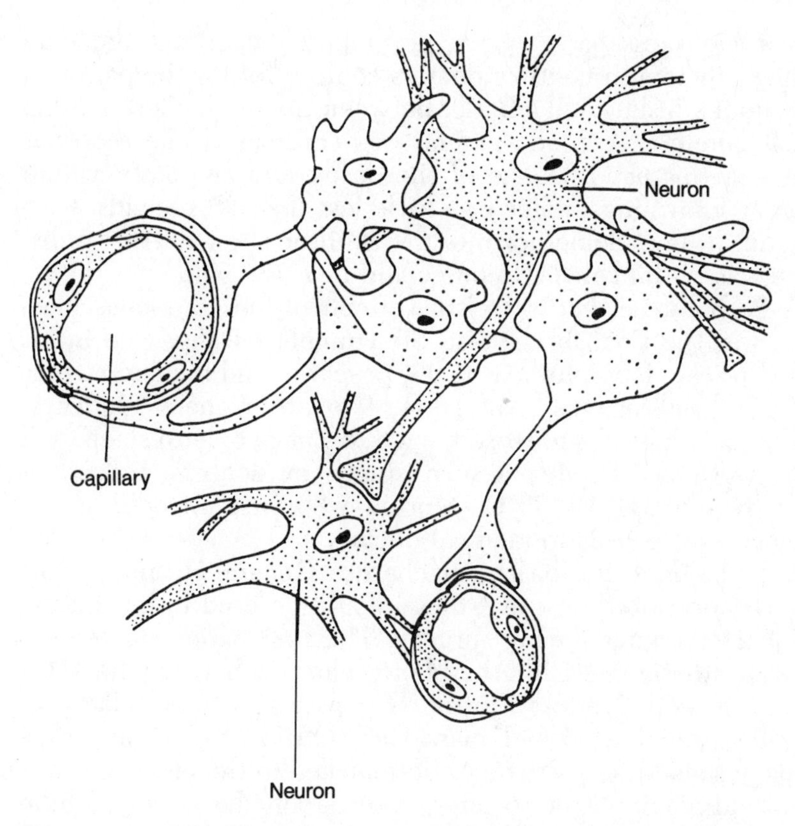

The nerves of the brain are protected by a barrier of sorts made up of tight-fitting capillary walls and special cells called astrocytes.

chemicals and toxic substances. Unfortunately, though, the one weakness in the system is that the blood-brain barrier cannot stop drugs that are soluble in oil or fat from passing through its gates. Oil-lipids make up the core of both the membranes of the capillary walls and the sheath of the astrocyte cells. Because of this, oil-soluble (or fat-soluble, its the same thing) drugs are able to enter the brain rapidly whether the brain wants them or not.

Still, the brain's weakness is medicine's gain. And no where else has that gain been greater than in the field of anesthesiology, the science of putting people to sleep. It is frightening to recall that it was just some one hundred and thirty years ago that surgeons were still operating on the human body without anesthesia. The drugs that could induce blissful unconsciousness hadn't been developed yet. About the only things nineteenth-century doctors could offer for the pain was a swig of whiskey and a bullet between the teeth. Certainly no great comfort. A good surgeon was considered one that was fast with his hands. Some of the best could amputate a limb from a squirming soldier in a little less than 25 seconds. Cutting into the abdomen or into the skull or chest to treat a disease was unthinkable. No one could bear the pain.

Perhaps the first to feel the effects of forced unconsciousness was the English chemist, Sir Humphry Davy, who in his small private laboratory in 1808, prepared and inhaled nitrous oxide, "laughing gas." By 1831, two other chemical drugs, ether and chloroform, were known to put people to sleep, and they were first used, not surprisingly, by dentists. Indeed, it was the dentist, W. T. G. Morton who first introduced the magic of anesthesia to medical doctors.

In 1846, at Boston's Massachusetts General Hospital, Morton demonstrated how anesthesia could be used to keep a patient unconscious during surgery. The first patient to receive this anesthesia was Gilbert Abbott, who was having a neck tumor removed. Dentist Morton had patient Abbott place his mouth around a special glass flask. Inside the flask was a sponge soaked with ether. Abbott sucked in the ether and was out cold a minute or so later, whereupon the surgeon, John Collins Warren, preceded to operate on the tumor.

In the fetal brain, nerve cells develop at an average rate of more than 250,000 per minute. At birth, a child's brain contains close to a trillion nerve cells. After that, few new nerve cells are added.

Bathed in the sterile, white glow of the operating room, and forced to stare at the cold, antiseptic instruments of surgery, the modern patient does not think of such historic events. But the modern anesthesiologist must. Armed with a legion of knock-out drugs, the specialist uses them carefully, gently, sparingly. Variations of as small as a fraction of one percent in the dosages of certain anesthesias can be lethal to the patient on the table. So the anesthesiologist draws on two sets of drugs. The first set is designed to work rapidly but briefly, to get the patient under as quickly as possible. The second is designed to keep the patient under for a longer period of time.

To begin, he injects into the vein a small amount of a member of the first set of drugs, sodium pentothal, to test the waters, so to speak, to see if the patient can tolerate more. Then, he adds an anesthetizing dose of the same substance. Sodium pentothal works quickly and induces unconsciousness in the patient in a matter of seconds. Extremely soluble in the fat-oil of cell membranes, it rapidly leaves the bloodstream and penetrates the brain.

But the duration of unconsciousness is brief. If it is not followed by another, longer lasting drug, the patient will wake up in 5 or 10 minutes after the injection. Sodium pentothal, being so soluble in oil, diffuses rapidly out of the brain, cutting short its action, and is transported by the circulatory system to the fatty deposits in the body. Hoarded by the fat cells of the body, the blood levels of sodium pentothal drop. And with little of it left in the brain, the patient wakes up.

The individual is still groggy, though, and in fact, will continue to be groggy for some time. The sleepy feeling is the result of the low levels of the drug remaining in the blood while it is slowly broken down by the liver. The drug is still present, but the concentration is not high enough to induce sleep or anesthesia.

Once the patient is under from the sodium pentothal, the anesthesiologist moves quickly before the patient wakes. He delivers the next drug: one that will put the patient lying on the table out for a longer time. The drug is usually in a gaseous or vaporized form and is breathed through the mouth. He may use nitrous oxide and supplement the gas with other anesthetic gases such as halothane or isoflurane. The vapors are extremely soluble in fat and flow past the lining of the lungs and into the bloodstream in a matter of seconds. There they flow into the brain and cause rapid unconsciousness.

The fat-soluble gases keep the patient asleep, but only if they are constantly fed into the lungs. For once the specialist shuts off the equipment supplying the gases, the concentration of the anesthetic in the lungs drops below the concentration of the anesthetic in the blood and the drugs readily and rapidly diffuse back into the lungs from the bloodstream and are then exhaled in the patient's expired air. Gradually, the patient awakens.

The situation where the patient wakes up during surgery with his innards still opened and exposed has been known to happen more than once. Some surgeons, then, might prefer to administer phenobarbital to their patients after an injection of sodium pentothal. Phenobarbital, a long-acting barbiturate, exists in the blood primarily in the water-soluble form and is only slightly soluble in muscle and fat. And unlike its faster-acting cousin, phenobarbital has some difficulty passing though the blood-brain barrier. Thus, it may take many minutes for phenobarbital to build up concentrations in the brain sufficient enough to induce sleep. Nevertheless, it circulates in the bloodstream in concentrations sufficient to keep the individual sedated or drowsy for several hours until it is either broken down by the liver or dumped out with the urine. Indeed, the great amount of the drug that persists in the bloodstream largely accounts for the hangover that lasts for several hours or even days after it has been administered into the body.

Bathed as it is in warm cerebro-spinal fluid and protected by a barrier of cells, the brain is like a child in a shielded womb. And yet, the brain is not immune from the effects of many drugs. But, then, neither is a woman's placenta, the organ which surrounds the real child. Indeed, one might be prone to believe that the placenta would be more immune to the invasion of drugs, considering a child's importance. But the placenta must provide the growing fetus with nutrients, and exchange waste products and hormones with the mother. This dependence of the fetus on the mother, though, places the fetus at the mercy of the placenta when drugs appear in the mother's blood.

Pregnant women are themselves exposed to a number of foreign substances. Cosmetics, household chemicals, fumes, prescription drugs, and drugs bought over the counter all get into a woman's circulatory system. Alcohol, for example, travels through both the placenta and the infant's blood-brain barrier. Fetal alcohol levels reach those of the drinking mother in about 15 minutes and can even be detected on the baby's breath at birth. Anesthesia, often used to relieve the pain of mothers in labor, also gets into the infant. Ten minutes after secobarbital is injected into the mother, the blood levels of the anesthetic are nearly identical in both mother and newborn baby. The infants delivered this way are lethargic.

Narcotics, barbiturates, DDT, even nicotine eventually get to meet the fetus. Back in 1979, Henry Sershen and Abel Lajtha, two neurochemists working at the Rockland Research Institute in New York City, attempted to find out not only how fast nicotine gets into the brain of fetuses, but also what this active ingredient of tobacco does when it gets there. Sershen already knew that nicotine mimics the action of the neurotransmitter acetylcholine in both the brain and the rest of the body. As such, it increases heart rate and blood flow, and triggers the release of adrenalin from the adrenal glands. It also stimulates the brain.

A person without oxygen will lose consciousness in about thirty seconds and will become comatose in about one minute.

To determine when and how the chemical was getting into the brain, Sershen and Lajtha first prepared nicotine molecules with attached radioactive carbon atoms to be used as tracers. Then, they injected the nicotine into pregnant rats and measured the levels of radioactivity over time in the blood and brains of both the fetuses and their mothers.

The first thing they found was that nicotine got into the brains of the rat mothers very rapidly, within a few seconds. But nicotine also left the brain quickly. Indeed, from their calculations, ninety percent of the nicotine had gone in and left the brain within five minutes. Nicotine was obviously fat-soluble, for it did not appear to be carried in by any molecular transport system at the brain.

The second thing they learned was that as fast as the nicotine entered the brains of the mothers, it entered the fetal brains, too. And, though the levels of nicotine in the baby rat brains were a quarter of what they were in the maternal brains, it was evidence enough to suggest that nicotine was able to pass through the placental barrier and the fetal blood-brain barrier almost as readily as water passes through them.

Ultimately, though, it is at the cell that nicotine or any other chemicals produce their effects on the body. Drugs, after all, simply interfere with or mimic the molecules in a cell that are there naturally. Imposter or ally, a chemical introduced into a cell will trigger a chain of events that will eventually result in a detectable response in the body. Thus, cocaine interferes with the action of the natural brain transmitter, norepinephrine. LSD hinders the activity of the brain transmitter, serotonin. Cyanide shuts off the energy producing mechanism in a cell by binding to a key protein in the system. Heroin drives into the same molecular parking space on the cell as the natural pain-killing substances called endorphins. Anesthetics dissolve in the lipids of the cell membranes of nerves and so block the transmission of nerve impulses. As a result of all of their actions, we either bubble over with glee, sink into deep depression, drop into unconsciousness, or die.

These, though, are exotic drugs. A more mundane one, but one that is just as potent in activating a cell, sits in the confines of a tea bag. With the exception, perhaps, of nicotine, caffeine is one of the most widely consumed drugs in the world. Indeed, as I sit and write these words, I am consuming it myself. Besides tea, though, caffeine is found in coffee, cola drinks, and chocolates, which explains its popularity.

The body burns calories fastest when exercise takes place within three hours after a meal.

Once it is consumed, caffeine is absorbed slowly into the bloodstream; only about 25 percent of it reaches the circulation sixty minutes after ingestion. Being primarily water-soluble, it has a difficult time crossing through the lining of the intestine. But once it flows into the bloodstream, it is distributed evenly throughout the body and crosses into the brain as well as into the womb.

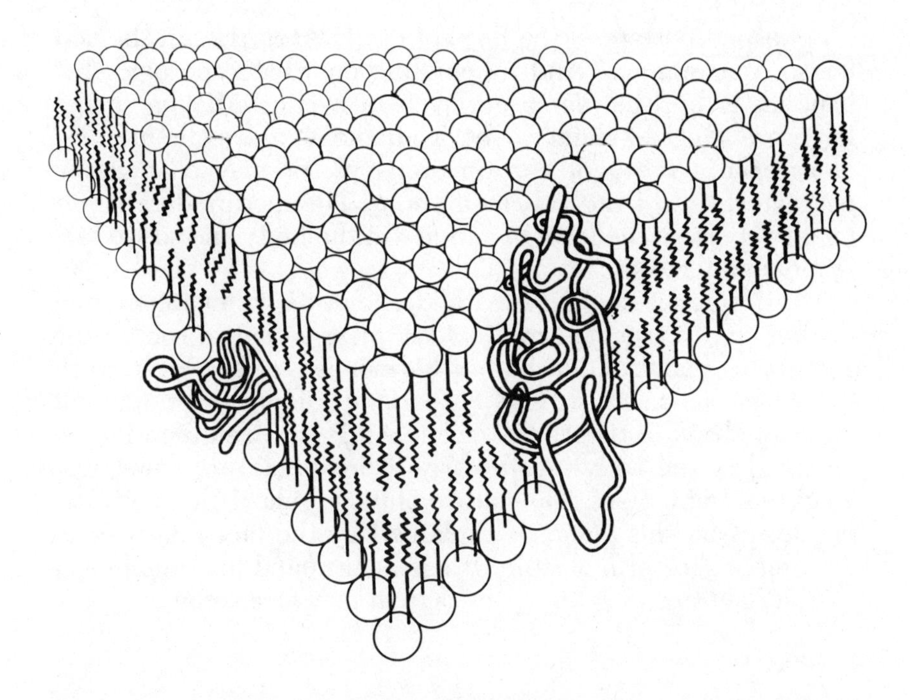

Many drugs must pass through the membrane of the cell. The cell membrane is composed of a sea of fatty molecules and several different proteins embedded within it.

Caffeine is a stimulant. It causes the heart to pump more blood and it also increases the flow of oxygen-rich blood to the heart. But, surprisingly, it decreases blood flow to the brain by constricting the blood vessels there. Nevertheless, caffeine, once in the brain, stimulates the cerebral cortex. We feel mentally alert and wide awake.

Caffeine does much of its work by worming its way into the cells of the body and inactivating a key enzyme there. The enzyme is called phosphodiesterase. Phosphodiesterase is responsible for deactivating a chemical known as cyclic adenosine monophosphate, or cyclic AMP. Thus, when caffeine shuts off phosphodiesterase, the level of cyclic AMP rises in the cell for several minutes. And, depending on what type of cell it happens to be, the increased levels of cyclic AMP may make for more glucose sugar to be available to the cell, may increase the activity of certain enzymes, or may change the characteristics of the cell membrane. The modified cells cumulatively raise the level of activity in the body and so we feel restless and awake.

Caffeine's association with cyclic AMP is an interesting one for, as it turns out, cyclic AMP plays a central role in the activity of many of the body's hormones. In fact, ever since the American scientist, Earl Sutherland, discovered this small molecule back in the 1940s, cyclic AMP has been found to be involved in the activity of nearly every organism, from pine trees to bumble bees. The nature of its popularity amongst the organisms on this planet is explained by the fact that it is an extremely ancient molecule. It has been found in a number of microorganisms. A billion years ago, it may have been used as a

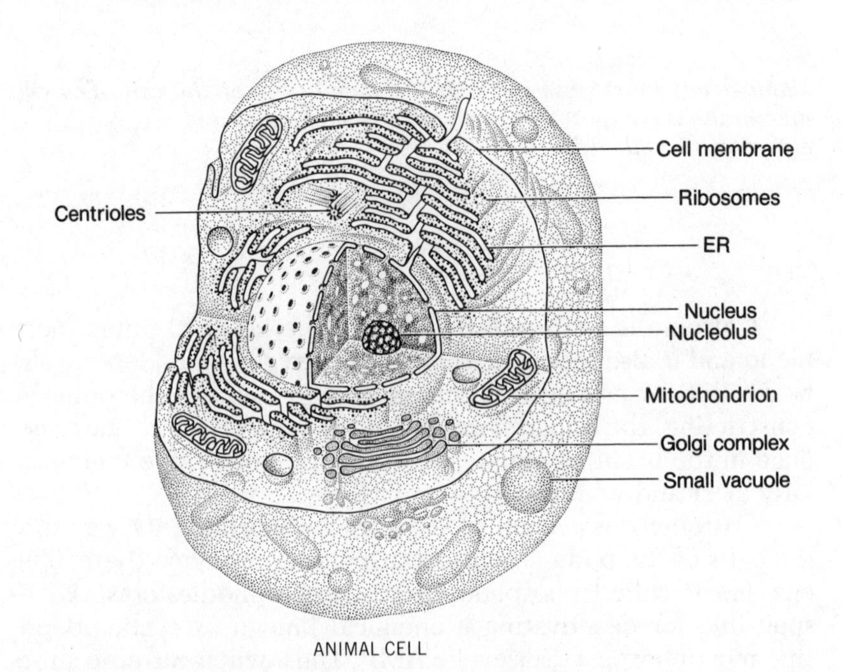

Ultimately, steroid hormones must do their business within the nucleus of the cell.

means of chemical communication among cells. Perhaps nowhere else is that function demonstrated better than in the activity of cellular slime molds.

Part cellular colony, part organism, the slime molds can be found on the forest floor eating the bacteria that decompose the leaves that have fallen to the ground. For much of their lives, slime molds live as solitary amoeba-like cells until they are threatened by starvation. When food grows scarce, millions of them stream together into a sluglike body. The chemical released during each cell's hunger pangs and used to attract these cells to come together to form an organism is cyclic AMP.

Somehow, for our purpose, cyclic AMP has become incorporated into our cells to be used as a kind of internal communicator. Within the cell, it shuttles about, delivering the message first introduced to the cell by hormones like adrenalin. In that regard, it has been called a "second messenger," the first messenger being the hormone itself.

Our cells require this second messenger in their bodies because many hormones, like adrenalin or ACTH, only deliver their message to the cell's surface. These hormones attach themselves to receptors on the cell membrane and, through these receptors, activate cyclic AMP inside. Cyclic AMP, in turn, triggers changes within the cell before this molecule itself is destroyed. Changes within the cell occur in a few seconds to a few minutes.

Adrenalin and a host of other hormones, cannot get into the cell because, like many drugs, they are not soluble in fat and therefore cannot pass through the cell's plasma membrane. Thus, they must rely on molecules like cyclic AMP to do their bidding within the cell. But, there are other hormones secreted in the body that have no use for a second messenger. They can readily pass into a cell by themselves.

Fat-soluble, the steroid hormones secreted by the adrenal cortex, by the testes, or by the ovaries, normally flow past the cell membrane and into the cell where they mobilize it to manufacture proteins. When, for example, the male sex hormone testosterone, secreted by the testes, gets into a muscle cell, muscle proteins are manufactured, and muscles get bigger.

The steroid hormones like testosterone, and estrogen (secreted by the ovaries) usually perform their jobs by attaching themselves to a receptor within the cell. Bound in this way, hormone and receptor move to the nucleus together and react with the DNA molecule there. The interaction is a powerful one, for within minutes after the steroid hormone has fraternized with the DNA, enzymes within the nucleus manufacture the blue-print for a protein. The blue-print, in the form of a messenger RNA molecule, diffuses back out into the cytoplasm of the cell where it is "read" by huge enzyme complexes which have the ultimate job of manufacturing the actual proteins in question.

The process, from hormone entrance to the start of protein manufacture by the cell, is not quick. Instructions must be copied and delivered. Proteins must be forged from scores, and sometimes even hundreds of amino acids. Then the proteins, too, must be delivered to places within the cell, their fate to remain or to be secreted out of the cell.

It takes four to eight seconds for something swallowed to move down into the stomach where it could remain as long as four hours.

When, for example, the adrenal hormone aldosterone enters a kidney cell, it binds to a receptor there and is brought into the nucleus. But only after some 45 minutes have gone by do the first proteins begin to appear in the cells. The 45 minute lag time between hormone entrance and protein manufacture is typical for most cells that welcome steroid hormones.

Slow or not, the effects of steroid hormones on the body are legion. The tall, muscular build, the deep voice, and the rough skin of males; the production of milk, the pubic hair, the monthly menstrual period of women, are the result of these steroid hormones. It would come as no surprise, then, that their action on the mind would be equally powerful.

In 1986, researchers Maria Majewska, Rochelle Schwartz, and Steve Paul of the National Institute of Mental Health (NIMH), and Neil Harrison and Jeffery Barker of the National Institute of Neurological and Communicative Disorders and Stroke, found that the metabolite of a steroid hormone in rats, THDOC, acts just like a barbiturate in its ability to inhibit or depress nerve cell activity.

Majewska and her co-workers discovered that the inhibitory steroid metabolite THDOC, mimics the effects of a naturally occurring inhibitory neurotransmitter in the brain called GABA (for gamma-amino-butyric acid). The effect is like that of the anesthetic phenobarbital, the chemical that depresses brain activity rather rapidly.

Oddly enough, the researchers also discovered that there are other steroids secreted by the body that are excitatory in nature, that block GABA's natural inhibitory effects. The interplay between the two different sets of steroid hormones, Majewska suggests, produces changes in an individual's mood and behavior.

Normally, the hypothalamus signals the pituitary gland beneath it to release the hormone ACTH. ACTH, in turn, triggers the outer portion of the adrenal gland (where the THDOC metabolite and other steroids are found) to secrete its hormones.

But during times of stress, or during depressive illness, the adrenals release greater amounts of steroid hormones into the bloodstream. Greater amounts of the barbiturate-like THDOC, then depresses nerve cell function in the brain. Steroids have the ability to cross both the blood-brain barrier and the membranes of cells rapidly. Interestingly, patients suffering from depression have been found to have high levels of cortisol in their blood. Cortisol is another steroid hormone secreted by the adrenal gland.

Assuming that what Majewska and her colleagues have found is true, it could mean that many of the mood changes associated with stress, anxiety, and depression might be related to the effects that steroids and their metabolites have on nerve cells in the brain.

One cause of chronic heartburn is a stomach that is slow to empty. Acid-soaked food builds up and then backs up into the esophagus causing the scorching irritation of heartburn.

When Napoleon Bonaparte was a young man, he was very thin and extremely ambitious. Sleep was a waste of time; only his work mattered.

But by the time the Emperor of France was in his forties, his mental and physical state had changed. His face became soft and feminine-looking, and his hips and chest had large deposits of fat on them. He slept a lot more than he used to, and his moods were, according to his closest followers, unpredictable. One moment he would bark out commands in a fit of rage; the next, he would collapse in his chair and whimper like a child. He complained that he was being poisoned by the British.

Napoleon himself noted the physical changes and said to Francesco Antommarchi, his personal physician, upon coming out of his room after an alcohol rub, "Look what lovely arms! What smooth white skin without a single hair! What rounded breasts! Any beauty would be proud of a bosom like mine!"

At dusk, on May the fifth, 1821, Napoleon died. He had expired after a long illness in his bed in Longwood House on the island of St. Helena in the South Atlantic. The next day, his personal physician performed the autopsy and confirmed his feminine characteristics. The Emperor's body was heavily covered with fat. There was scarcely a hair on his body, and his penis and testes were small and resembled the *mons veneris* in a woman. It seems now that he suffered from adiposogenital dystrophy, a disease of the brain's hypothalamus. The disease causes deranged pituitary function and produces abnormal adrenal gland secretions. It is possible that the steroids from his adrenals caused his sleepiness and erratic moods.

But there is another theory for his behavior. In 1961, Swedish dentist Sten Forshufvud, forensic scientist Hamilton Smith, and Anders Wassen, analyzed a sample of the emperor's hair that had been clipped from his head the day after his death. Hair often incorporates certain chemicals from the body. In it, the three scientists found traces of the poison arsenic. Given in small doses over a period of days, arsenic causes lethargy and fatigue, nausea, and muscle tenderness in its victims, and eventually, death.

Either way, he died an unhappy man.

FIVE

THE BODY IN HOURS

Gamblers will bet on anything—football games, political races, fights. If the odds are good, they'll even bet on your life. But don't bet against them. You'd lose.

To the Las Vegas odds makers, betting on the human body is a good risk. Over the years, chronobiologists—those scientists who study the body clock—have collected an encyclopedia of events that predictably rise and fall in the human body over the course of a solar day. The information would make any gambler rich.

The odds are good that we humans are more likely to be born, drop dead from a heart attack, or fall asleep during one part of the day than during another. Hormone levels, body temperature, blood pressure, heart rate, cell division, even the strength of a handshake, have their predictable peaks and troughs.

The favorable edge comes from the reliance of the human body on its own rhythms—ancient, internal clocks of sorts that keep our bodily activities in check with the 24-hour, light/dark, solar day. Actually, the cycles rarely ever match an exact solar day. They simply approximate it. Thirty years ago, University of Minnesota researcher, Franz Halberg, took note of this fact, and coined the term, *circadian,* from an alloy of two Latin words: *circa* (about), and *dies* (a day).

The peaks and troughs in the body rhythms are controlled internally, but they are continuously reset to an exact 24-hour day by the light cycle to which the body is exposed. As a result, jet setters, who travel across many time zones, or shift workers, who change their routine from day to night, can suffer from an internal cycle that drifts slightly from the external one. The consequences over time can be diarrhea, constipation, malaise, and even ulcers, as the body slowly re-adjusts to the new day/night schedule.

The rise and fall of body activity or even birth and death, is not exclusive to humans. Indeed, circadian rhythms are found in just about every creature on the planet, from bread molds and bacteria, to birds, bees, and bears. The universality of this phenomenon is necessary for the survival of the species. By incorporating an internal program of the movements of the sun into their cells, organisms are able to prepare their behavior or their physiology *in advance* of the changes in the environment. This internal calendar is so critical to life, that it is capable of oscillating alone if the organism is kept in isolation. It is so important that it is inherited, never learned.

Fruit flies, for example, emerge from their immature wormlike pupae stage into their adult flying stage about two hours before noon. But the escape is not triggered by the movements of the sun per se. The flies emerge before noon because the instructions to do so are burned into their genes.

Raise fifteen generations of fruit flies under continuous dim light which, in effect, removes the natural rhythmic signals of light and dark, and the sixteenth generation of flies will still emerge from their pupae stage on schedule—about two hours before noon each day.

The "about-a-day" cycles were first observed in plants as far back as 1729, when French astronomer Jean de Mairan noticed that a heliotrope plant always opened its leaves in the morning and closed them at night. The slow dance of the heliotrope's leaves were paired to the plant's internal circadian rhythm and required no movement of the sun to initiate it. As de Mairan later demonstrated, the leaves would open and close even when the plant was placed in constant darkness for a day.

Nicotine, a component of tobacco smoke, gets into the body very rapidly. With one puff on a cigarette, nicotine reaches the brain in 7 seconds. This is several seconds faster than it takes for alcohol to get into the brain. Even heroin, when it is injected into the arm, takes twice as long as it does for nicotine to reach the brain.

The history of demonstrating the innate character of plant and animal circadian rhythms was fairly straightforward compared to what it took to show that humans themselves possessed internal cycles. In 1657, Galileo's colleague, Sanctorius, built an enormous balance with an immense platform on which he constructed an entire room, completely furnished with a table, a chair, and a bed. He erected this bizarre contraption so that he could record his daily or weekly changes in body weight while he lived on this tray for several consecutive months. Though it was a rather crude device, and a bit unsteady (it jiggled when he moved about), Sanctorius actually discovered that he weighed less during one hour of the day than another. He attributed this circadian rhythm in body weight to a loss of water through perspiration.

Nearly two hundred years later, the English researcher, Briton J. Davy, recorded his body temperature over the course of several days and even over a period of a year. Though, it must have been a rather painful procedure after a time (the thermometer *wasn't* stuck in his mouth), the rhythm-and-bruise method paid off. The hourly change in his body temperature was not related to his jaunts on horseback or to his running, nor was it altered by the temperature outside. It had a truly innate character.

Davy's discovery, that body temperature rises and falls throughout the day, has been refined by current-day scientists. As more modern researchers have found, body temperature doesn't remain around 98.6 degrees throughout the day. It rises to a high of about 99.0 degrees by five or six in the evening, and gradually drops to a low of 97.0 degrees by three to four in the morning. Even our heartbeat is not constant over the course of a day. It may vary by as much as twenty to thirty beats per minute over the stretch of 24 hours. Hormone levels fluctuate. Mood and memory swing up and down. Muscle contraction varies throughout the day. A handgrip, for instance, is strongest around dinner time (6:00 P.M.) and weakest at 3:00 A.M. Virtually *every* body function shows a circadian rhythm.

A single cancer cell divides, on average, only once every hundred days. At that rate, however, owing to exponential growth—one cell dividing into two, two into four, four cells into eight, and so on—the aberrant cell may take eight years to form a pea-sized lump; two years later, the tumor will be about the size of a cantaloupe mellon and weigh more than a pound.

Unnoticed, even the cells making up the inner lining of the skin and digestive tract, speed up their metabolic activity around midnight only to slow down during the height of the day. The normal rhythmic progression of cell activity is so predictable that it has been used to identify what is and what is not normal in the body. Rogue tumor cells of a cancerous breast follow their own rhythm, speeding up their activities every 20 hours instead of the normal 24. The unusual cyclic changes in the surface temperature of a cancerous breast (due to the activity of the cells) offer a means of detecting their once hidden behavior.

Events in our mouths also change throughout the day. For

example, while more than a quart of saliva is secreted by the glands in the mouth each day, very little of it is secreted at night. The bacteria that grow on our teeth somehow know this. They divide more rapidly when wc are asleep. Their frantic activity does not go completely unnoticed. We wake up with a slimy film on our teeth, and usually with bad breath.

Even the body's enzymes cut and splice faster during one part of the day than another. Many of these enzymes keep strange hours. The enzymes that break down the toxins and drugs we ingest, are at their speediest around two in the morning and at their slowest some twelve hours later.

Oddly, the liver enzyme that has the responsibility of destroying the alcohol we imbibe, has a different schedule in men than it does in women. For men, the enzyme, alcohol dehydrogenase, is at its fastest at around eight in the morning; for women, for some strange reason, this enzyme's speediest performance occurs around 3:00 A.M. That kind of sexist biochemistry may be one of the reasons why a woman can often drink a man under the table in the wee hours of the morning, and why a lot of men (hoping that they can get their dates drunk) often go home intoxicated, disappointed, and alone.

Circadian Sickness

The marriage of the human body to the sun is an intimate and binding contract, one that is obeyed by the body in sickness as well as in health. Indeed, human ailments follow the rhythms of the sun, too, though we once assumed that disease and sickness occurred at all hours of the day. Actually, they are more severe, or are more likely to occur during one part of the day than another.

The sneezing, stuffy nose and red, itchy eyes of hay fever are at their most bothersome in the morning before breakfast, and don't subside until lunch. Asthma and dyspnea (labored breathing) are more likely to attack during the night than during the day, an observation that has been known for centuries. "The evil (dyspnea) is much worse in sleep . . . ," the Greek physician, Aretaeus, wrote in the third century A.D.

Asthma, for one, occurs more often at night than during the day not simply because there is more pollen or dust in the air in the evening than during the day, but because of the fluctuations in the body's immune system. When in 1971, French physician Alain Reinberg, and Texas environmental scientist Michael Smolensky, found that the airways of asthmatic patients were more sensitive to the actions of inflammation chemicals such as histamine during the night than during the day, their patients were isolated in an air-conditioned (and air-filtered) room, isolated completely from the outside environment.

Then in 1977, Reinberg, Smolensky, and their colleague Pierre Gervais went one step further. They made up an aerosol spray filled with house dust and asked their patients to inhale the dust during various times of the day. In this way, they could control the dose of house dust their patients would inhale. Once again, the scientists discovered that the asthmatic patients were much more sensitive to house dust around midnight than they were at any other time during the day. The three scientists demonstrated that the reactions to pollen and other allergens are more likely due to an internal circadian rhythm of the sufferer's immune system than from the amount of, say, dust or pollen in the air.

If the body's sensitivity to pollen has a daily rhythm, so too does pain. Indeed, the threshold for pain is lower at night than it is during the afternoon. A toothache, for instance, is more painful during the early morning hours between 3:00 A.M. and sunrise than it is during any other part of the day. The onset of migraine and muscle headaches, for some as yet unknown reason, also begin in the early hours of the morning. A friend of mine wakes up with them.

Surrounded as we are by a host of viruses and bacteria, it is not surprising to learn that our body chemistry displays a circadian rhythm in its interaction with these microscopic agitators. In fact, the onset of fever resulting from bacterial infection occurs mainly during the morning between the hours of 5:00 A.M. and noon. On the other hand, the onset of fever from a viral infection, occurs mainly during the later afternoon and evening between 2:00 P.M. and 10:00 P.M.

That fevers from colds and influenza, inflammations and infected sores run in cycles makes sense when one looks at the activity of the body's immune system, the defensive organization designed to combat these microscopic pests. It, too, is more active during one hour than another. White blood cells, for instance, play a large role in the immune response. They are more active during the evening than they are during the day. Yet, try to explain the circadian nature of heart attacks, angina, or stroke, and the picture becomes far muddier.

In 1984, James Muller of Harvard Medical School was looking through the files of nearly 3000 patients who had suffered heart attacks. Flipping through the pages and pages of data, he suddenly began to see a pattern. To his great astonishment, most of the patients tended to have their attacks around nine in the morning. Indeed, the chances of a heart attack striking at 9:00 A.M. was nearly three times greater than at 11:00 P.M. Heart attacks were *least* likely to occur around 9:00 at night.

Excited, Muller got in touch with Thomas Robertson and John Marler of the National Institutes of Health, and asked them to run a computer search to find out at what hours of the day strokes were most likely to occur. To everyone's surprise, the data indicated that strokes, too, tended to occur around 9:00 A.M. They were least likely to happen between three and four in the morning.

About a quart of blood flows through the brain every minute.

Meanwhile, Paul Ludmer, with Cardiovascular Consultants in Oakland, California, looking over the death certificates of persons who had died of sudden death in Massachusetts in 1983, and Michael Rocco of Harvard, perusing the data on the timing of transient ischemic attacks (where the heart muscle is deprived of blood), discovered the same thing. Most of the incidents occurred in the morning, around 9:00 A.M. This is mysterious indeed. Yet, its underlining cause may be due more to the circadian rhythms of our body's hormones and biochemistry than to anything else.

Heart attacks occur when a clump of blood clots a coronary blood vessel supplying the muscles of the heart. Oddly enough, there is a circadian rhythm in the aggregation of platelets, the small, flakelike fragments of bone-marrow cells which cause blood cells to clump together. There is also a daily cycle in their response to heparin, a natural chemical of the body which prevents clotting. Though no one yet knows why, platelets are much gooier in the morning, and thus more prone to form clots, than at other times of the day.

Blood clotting, however, doesn't explain why heart ischemia occurs in the morning. Ischemia is often caused by spasms of the coronary vessels. By coincidence, the levels of adrenalin in the blood tend to increase in the morning. Adrenalin causes coronary blood vessels to contract. This still doesn't reveal why strokes tend to occur around 9:00 A.M. As Marler told a reporter from *Science:* "We just don't know what causes strokes to happen, period, let alone why they happen at a particular time of day." The timing of death from these severe cardiovascular incidents is still a mystery.

". . . To die: To sleep . . . "

Centuries ago, sleep was confused with death. For the early Germanic tribes during the time of the Roman Emperors, death and sleep were inseparable. Even as late as the Elizabethan period of English history, death and sleep were sometimes mistaken. William Shakespeare ended his play *Romeo and Juliet* by making the heroine take a drug that knocked her out. Her deathlike sleep not only fooled her family, it fooled her lover, who killed himself. Centuries later, author Raymond Chandler wrote of "sleeping the Big Sleep. . . ."

The need to sleep is irresistible; most of us spend, on average, about seven or eight hours a night—a third of our lives—asleep. Some people may require less; some, more. Napoleon, for one, needed no more than four hours of sleep. Albert Einstein, on the other hand, spent ten hours of night in bed. But no one can live without it.

Deprived of sleep for only a few hours, we feel listless. Deprived of it for a few days, we feel awful. Sleep deprivation has been used throughout history as a form of torture or punishment. The ancient Romans used the *Tormentum vigiliae*, or waking torture, to slowly kill their prisoners. During the Middle Ages, the *Tortura insomniae,* or insomnia torture, was used to extract confessions. According to the *Guinness Book of World Records,* the longest anyone has ever been able to repel sleep's tantalizing lure has been about twelve days. After that, sleep claims its victim.

Yet, its purpose still eludes us. Nearly universal in animals—pigeons, dolphins, rats, even moths and flies exhibit a kind of sleep behavior—its ancient circadian periodicity may have been designed for protection. By imposing a period of inactivity, it restricted an organism to its hiding place, safe from predators. Yet, a sleeping animal is easier to sneak up on than an awake one.

Sleep may also have evolved as a period to restore the body's worn out parts. Here, too, it has done a poor job. While an organism consumes less energy asleep than awake—even its body temperature is lower during this period—there is no increase in the production of protein, a necessary component of tissue, during this time. In fact, there is a decrease in the rate of protein construction during slumber.

If sleep were the result of a system shutdown, its purpose would make some sense; a necessary recess to give the brain a rest. But it isn't. Sleep is an active process. Stimulate the reticular formation—the cluster of nerve cells at the base of the brain—with an electrical jolt and an animal will fall asleep. Far from being dormant, a sleeping brain is busy and animated.

Half the Valium ingested is eliminated by the body in 20 to 90 hours.

If we do not know why we sleep, we do have some idea as to why we fall asleep at night. The reason may have more to do with a certain activity in our brains, than whether we are simply just tired. The pineal gland, a small, white, pine cone-shaped structure in the middle of the brain—a structure that displays a circadian rhythm—makes and secretes far more of the hormone, melatonin, at night than it does during the day.

It is so potent, in fact, that when this hormone was administered to adult male volunteers during the daytime, it made the men sleepy. Children, who produce more of it at night than do adults, sleep more than their parents because of its potent actions.

Modern sleep research got its boost in the 1940s when Nathaniel Kleitman, working in Chicago, developed an interest in the slow, sweeping eye movements that occurred when his human subjects fell asleep. Too busy with his own work, he handed the project over to his young graduate student, Eugene Aserinsky, who recorded the electrical activity of the muscles in the eye socket as their coordinated contractions moved the eye up and down, back and forth. With this electrooculogram (EOG), Aserinsky soon discovered that long after his human subjects had fallen asleep, the needle of the EOG would vibrate violently on the paper. The shivering needle matched the brisk eye movements occurring beneath their lids. The motion of the eyes was significant. When they were abruptly awakened, the sleeping subjects reported that they had been dreaming. This newly discovered stage of slumber was given the term, rapid eye movement, or REM sleep. REM sleep was an exciting find. But its discovery just added one more bone to sleep's complex skeleton. Years before, researchers had found four other phases of sleep (labeled simply 1,2,3, and 4), each with its own peculiar characteristic. Now, because of REM's prominence, their importance waned. Collectively, they were given the label of non-REM sleep. Important or not, together the alternating sequence of REM and non-REM makes up the four or five cycles, each lasting about an hour and thirty minutes, that fill the routine of a full night's sleep. The hours are busy ones.

Sleep's Autograph

The sleeping brain does not yield its mysteries willingly. Its secrets must be drawn out using the most sensitive of electrical equipment. Electrodes must be attached to the skull to record on paper the electroencephalographic, or EEG patterns of the

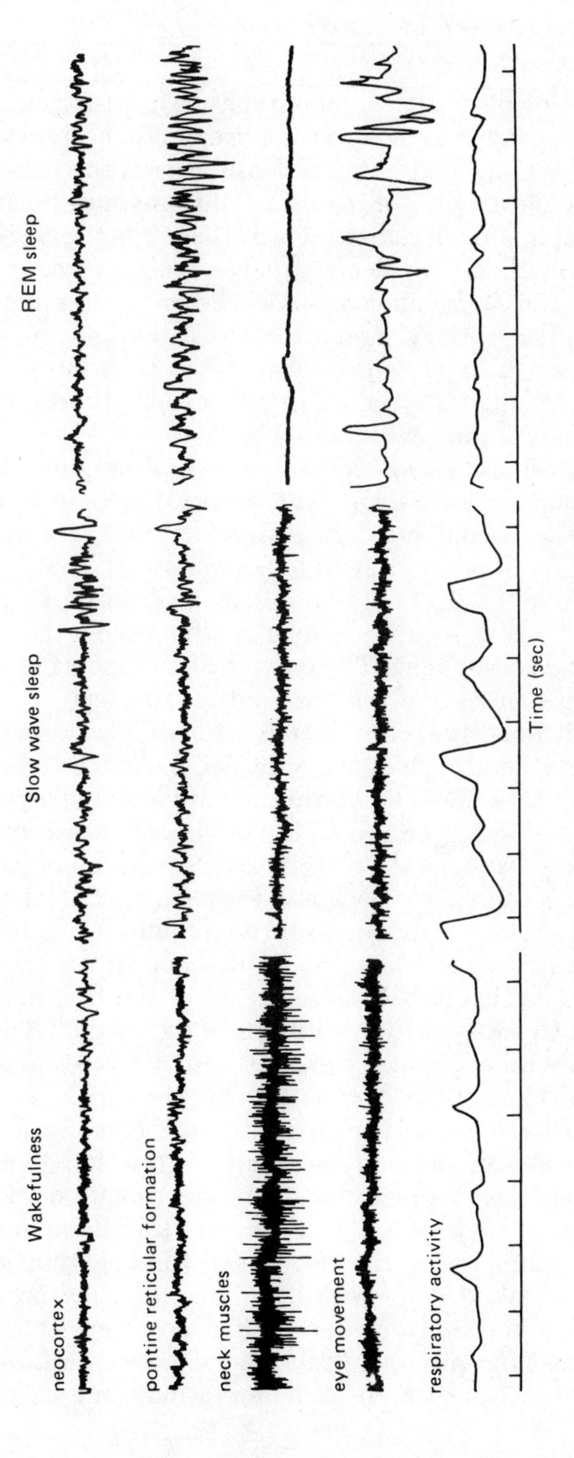

The signature of sleep changes throughout the night.

brain, violent electrical autographs that arise from the activity of a billion different nerves. Wires must be pressed onto the skin of the jaw and onto the flesh around the eyes to measure the fine movements of muscles. All these must be attached to a polygraph, a device that records these electrical signals on paper. Only then does sleep reveal its true nature.

To watch the movements of the polygraph needles as they scratch their jittery signatures across the slow current of moving paper, is to spy on a world that is both intimately familiar yet totally alien, for the journey is a roller-coaster ride through the inside of our own minds.

As we fall asleep, several events occur. Body temperature and blood pressure drop. At the same time, our rate of breathing and even our heart rate slows as the brain makes adjustments for our journey into slumber. At first, the needles recording the EEG vibrate back and forth in rapid, short strokes. These so-called alpha rhythms show that the brain is still relaxed but alert. The irregular movements of the eyes and the tense muscles of the jaw confirm this fact.

But roughly three minutes later, the foundation on which our alert mind rests, gives way. We sink into the netherworld of sleep. The EEG alpha rhythm suddenly changes. The once rough scribbling of the EEG now becomes a timid, yet rapid and irregular scrawl. The muscles of the body relax slightly. The eyes roll slowly back and forth inside their homes. This is stage 1. It lasts only one to three minutes.

As this stage collapses, we descend to the second stage of sleep. The timorous EEG scratches of the first phase give way to slightly larger EEG vibrations of the second. They are overlapped with occasional bursts of activity waves known as sleep spindles. Sometimes the pattern is interrupted by the appearance of large, sweeping waves called K-complexes. Meanwhile, the muscles of the body are only slightly less tense than they were when we were awake. The eyes remain motionless.

We remain at stage 2 for ten to fifteen minutes, then slowly submerge to the third. The EEG rhythms of the brain are now called delta waves. Large, slow, and lazy, they indicate that the body has entered deep sleep. It is during this stage that the pituitary gland at the base of the brain releases large bursts of growth hormone, a molecule that coordinates the body's metabolism. This is also the place where, during puberty, most of the male sex hormone, testosterone, and LH, a chemical required in the female for her monthly release of eggs, are secreted into the blood. Yet, despite its importance, we stay here only for some five to fifteen minutes.

Down we sink to sleep's deepest pit, stage 4. Here in this benthos of slumber, we stay for thirty minutes, our muscles and eyes relatively still. Then, we travel up to stage 3, remaining there for only ten minutes until the psychic elevator moves us up to stage 2 for another ten.

For about an hour and twenty minutes our mind is a corked bottle riding the slow waves of sleep. Then, suddenly, a door opens. The visions are chaotic; the scenes, bizarre, bright, and colorful. We are in REM sleep and we are dreaming.

The psychic episodes match the frantic movements of the polygraph needles. The signals coming from the muscles of the eyes indicate that they are moving about rapidly. The EEG pattern resembles the nervous scribbling of stage one sleep characteristic of mental alertness, yet the muscles of the rest of the body are limp and useless. Despite this, breathing and pulse rate quicken, and blood pressure rises. In males, the penis becomes erect, though men are not necessarily dreaming about a sexual episode. Even infants have erections during this period.

Like sleep itself, REM sleep is not a uniform event. It alternates between two smaller events called REM-M and REM-Q, named for the characteristic motions of our eyes—moving (M), and quiescent (Q). In REM-M, our eyes follow the activity on the dream stage. As the actors we have created move about, so do our eyes track them. Yet the muscles of our legs do not chase these images. Except for a few muscle twitches, the brain has inhibited their movement.

As one dream ends, we enter REM-Q, a kind of mystical intermission between scenes. The eyes remain still, but the body tends to move about. If we were to be awakened now, we would report that our dream had just finished and we would probably remember it the next morning. If left undisturbed, we would dive back into REM-M, once again.

It is now 90 minutes after we have first fallen asleep and a new cycle begins. Torn from our dreams, we travel back up to stage 2, to dwell there for a longer period of time than before, about 20 to 30 minutes. Then its back down to stage 3 (for 5 to 10 minutes), and deeper still to stage 4. We visit this island for only about ten minutes. Then, up again to stage 3 (five minutes), and to stage 2 (five minutes), only to nose dive back into the warm, fluid sea of REM sleep. Here, we extend our visit for almost fifteen minutes.

This second cycle has taken as long as the one before (an hour and a half), yet something interesting is happening. The stages of deep sleep, stages 3 and 4, begin to disappear. Indeed, by the third 90 minute cycle, only REM and stage 2 are left to toss the sleep ball equally amongst themselves for our journey through the night. Until we awake.

Flying trees and swimming office chairs are the stuff of dreams. As such, their bizarre presence is wholly natural. Yet, the reason for REM sleep—a state we all visit for nearly two hours a night, and the mother of these hallucinations—is far more perplexing. Supposedly a primitive form of sleep—cats and other mammals spend a large amount of time before birth in a conditions similar to it—REM sleep may have evolved as a program to rehearse such genetically determined functions as instincts and sexual behavior without having to physically act them out. It may also have evolved as a period of time in which to sort and file the events of the day.

After the age of forty, one loses about 1000 nerve cells a day. Even so, with a trillion nerve cells in the brain, such a loss of cells usually has little effect on memory.

Whatever sleep's real purpose, our species spends a lot of the solar day in its bosom. In the early months of life, infants sleep most of the day away. Oddly, the EEG patterns of their REM sleep look surprisingly similar to the EEG patterns when they are awake.

The elderly tend to sleep more during the day than at night, when most people do. When they do fall asleep, it is often fitful—they wake up repeatedly—and erratic. They tend to spend less time in deep sleep than younger adults do.

Yet, the percentage of REM sleep in both the old and the young remains relatively constant.

The fact that it invades our lives so thoroughly attests to sleep's power and to its importance. It plays a vital role even during puberty, when many hormones are primarily released during sleep. Yet, despite its prominence, it is still a slave to our circadian rhythms. Its Master, though, has been found.

The Master Clock

In the 1920s, Curt Richter, a psychobiologist at Johns Hopkins University in Maryland, noticed that the rats in his laboratory still persisted to eat at night and sleep during the afternoon despite the fact that he had kept the room lights off for several days. They were obviously displaying circadian rhythms but, he pondered, where were these rhythms eminating from? Where was the Master clock?

He got his answer after a lengthy process of elimination. He sliced into a bit of brain tissue, saw that the animals still displayed their rest/activity rhythm, then sliced into another section of the nerve pudding.

This went on, year after year until, one day he sliced into the rats' hypothalamus near the base of the brain. For a day or so after the operation, he left his rats alone. But, when he later checked up on them, he was amazed to find that his rats did not display any of their normal circadian behaviors. They ate at all times of the day and night, and slept erradically.

When he later examined the brains of these rats, he discovered that the region of the brain that he had damaged contained nothing more than a minuscule cluster of nerve cells located directly above the crossing of the optic nerves (called the chiasma opticum), behind the back of the eyes and at the base of the brain, making up part of the hypothalamus. Because of its location, it was given the name, suprachiasmatic nucleus, or SCN.

For years, though, Richter's work went practically unnoticed until 1972, when two experimental psychologists, Fred Stephan and Irving Zucker, both from the University of California, sliced into the SCN, ripping though the fragile nerve cells and disrupting their connections with the rest of the brain, and discovered once again that the rats lost their circadian rhythms. The rats ate, slept, and ran around during all parts of the day. It was very unusual.

More importantly, when Stephan and Zucker severed the optic nerves of their rats, essentially blinding the animals, their biological rhythms still continued. This was a very important discovery indeed. It meant that the SCN had its own private neural link with the retina independent of the optic nerve. It had an uninterrupted connection with the sun.

The University of California team proved that the SCN got its orders from the movements of the sun, but that was about it. There was still no proof that the rhythms of the SCN could direct the actions of the body by itself—alone. That answer was discovered only a few years later by two different research groups.

In 1977, William Schwartz and Harold Gainer of the National Institutes of Health, were using a delicate biochemical technique to measure the metabolic activity of nerve cells in the brain of rats when their experiment revealed that the activity of the cells of the SCN had a circadian rhythm all their own. The cells were active during the day, but inactive at night.

Escherichia coli (more commonly known as *E. coli*), a bacterium that lives in the human gut, divides once every 40 minutes. At that rate, at the end of 24 hours, there would be some 17 million new bacteria in the gut.

Encouraged by this discovery, two Japanese scientists, Shin-ichi Inouye and Hiroshi Kawamura working at the Mitsubichi Institute near Tokyo, set out to prove that the SCN was not getting its instructions from other regions of the brain, but was in fact directing the rhythms of these areas by its own actions.

To show this, they sunk several, thin electrodes into the rat's brain, including one in the SCN, and looked at the electrical signals emanating from them. At first, the discovery was disappointing. All of the regions of the brain showed circadian rhythms in their activity. But, then, they sliced through the SCN itself, severing its nervous connections with the rest of the brain. The results were dramatic. Every region of the brain, save the SCN, no longer showed circadian rhythms in their activity. The two Japanese workers had finally confirmed that the SCN was indeed the "Master clock" Curt Richter had proposed some 60 years before.

The possession of a clock to coordinate the circadian activities of the human body is so vital to our survival, that it would make evolutionary sense to have it up and running and tuned to the sun as soon as we are born. In rats and squirrel monkeys, at least, that is exactly what happens. The SCN clock in these animals is up and running days before its connections with the retina have been made, and before the body itself expresses circadian rhythms. Newborn animals, even newborn humans, do not immediately express daily enzymatic and hormonal rhythms or regular sleep-wake cycles. So, for a while anyway, the SCN timepiece ticks but the body doesn't hear it.

The reason for its lonely vigil is that the SCN clock is set not when these animals are born, but when they are still in the womb. The timing of the fetal clock is fixed to the movement of the sun by the mother. As soon as they are born, and while their brain is still maturing, the mother continues to influence the ticking clocks of her offspring.

How the mother sets the SCN of her children to the light/dark cycles of the outside world even before they leave the womb, had been a mystery for years. Then, as late as 1983, the question was solved by Steven Reppert and William Schwartz of the Massachusetts General Hospital. In a cleverly designed experiment, they injected pregnant rats with radioactively labelled deoxyglucose several times throughout the day. Deoxyglucose is an analog of glucose sugar that is picked up by the brain and hoarded without being fully broken down. It thus serves as a marker for cell metabolic activity; the more active the region, the more isotope it will have. Moreover, the isotope travels through the placenta and lodges in fetal brains as well as the brains of their mothers. This way, the two researchers could measure the activity of the SCN of both the mother and her fetuses.

Reppert and Schwartz discovered that a circadian rhythm of activity develops in the rat fetuses' SCN three days before they are born, shortly after the cells that produce the SCN are formed. Rats have a gestation period of about 21 days. The question was: How did it happen?

Somehow the fetuses knew what time of the day it was even when they could not possibly have seen it. Yet, they were not getting the information directly from sunlight penetrating through their mother's abdomen and into the womb. The two Massachusetts researchers proved this by first blinding the mother rats, then reversing the day/night cycle in the lab. In this way, the blinded mothers would still have their clock running on the old cycle. Then, they waited for the fetuses to make a choice: their mother's old clock, or the reversed light/dark cycle in the lab. Surprisingly, the immature rats chose the out-of-phase circadian rhythms of their mother's. Moreover, when Reppert, and later Fred Davis, of the University of Virginia, removed the mother's SCN early in her pregnancy, destroying her own circadian rhythms, the fetal clocks continued to run but were out of sync with the outside world. Fetal squirrel monkeys show the same behavior.

The fetuses get their information indirectly from their mother. The theory is that sunlight falls on the mother's retina and resets the rhythms of her SCN first. Then a chemical signal (no one knows what) travels to the SCN of the fetus and resets it.

What this may mean for our own species is far from clear. Humans do have an SCN that is up and running by the seventh month of gestation. But, if the clock is functioning in human fetuses in the same way that it functions in rats and squirrel monkeys, then we must think about the consequences for our young. Shift work and plane trips across a number of time zones are known to screw up an adult's internal circadian rhythms. But what effect do these events have on a fetus lying inside the womb of a jet-setting mother, or a woman who shifts to working nights?

We do not yet know whether the clocks of human fetuses work in the same way that rats do. But if they do, then an abnormal signal from the mother's SCN may have serious repercussions in the timing of normal birth. Babies born in the morning usually arrive healthy and strong. Afternoon babies, especially those born between the hours of 2:00 P.M. and 4:00 P.M., are more likely to have medical complications. Interestingly, the incidence of stillbirth is slightly greater in the early afternoon than it is at other times of the day or night.

Midwives were frequently employed to help deliver babies. (1554 A.D.)

Birth Time

Whether or not the SCN is responsible for the timing of the trials and tribulations of nocturnal labor and morning birth in our species isn't known. Certainly, millions of years ago, there must have been a reason for the synchronization of these events with the dawn. But, if their timing is an ancient process, so too are the events which lead up to the birth of a child.

Despite its primitive and obvious necessity, birth has never been quick and easy. Before the advent of modern medicine, women endured the contractions of labor as long as the birth went normally. But every mother trembled at the thought of the baby lying within her womb in an awkward position. The pain and suffering from a protracted labor could last for days until the woman died with her child still unborn. No wonder women anticipated childbirth with dread. "Pray for me that I may bring forth this child," The Countess of Eglinton wrote to a friend in 1616, "and live to deserve your innumerable kindnesses."

To speed things up, relatives would offer strange herbal potions and spells. In medieval Europe, pregnant mothers were sometimes whipped while in labor so that they would give up their child faster, the theory being that the pain relieved the tension holding the child back. But (according to Martin Luther), when the Empress of Germany went into labor no one dared whip her. Instead, she had two dozen men brought into her lying-in chamber and had them lashed in front of her.

During the final stages of labor, the midwife, a woman who was part doctor, part medicine-man, often took charge. Usually, she let nature take its course. Occasionally midwives gave the mother hard liquor or eggs and broth if she seemed weak. They also examined her birth canal to check the progress of labor.

In seventeenth-century Europe, few mothers gave birth by lying flat in bed. Mostly, they squatted on a stool, or sat on another woman's lap, or knelt on a clean board. As the child emerged from the birth canal, the midwives would catch it, tie and cut the umbilical cord, and pull out the placenta, or afterbirth, that still remained in the womb. If the process took too long, they applied a magnet or a horseshoe to the mother's vagina to draw out the child. Sometimes, they would deliver the child feet first. Those women who recovered from this long ordeal unscathed were fortunate. "It hath pleased God to give me a safe delivery of a living daughter," Lady Christian Lindsay wrote in 1666, "though I had fears of the death of both her and myself before I was brought to bed, I was so extremely sick."

Whatever its social history, human birth as a biological event has traversed nearly the same road since our species first evolved. The slow, often half-a-day-long journey of labor and delivery is nature's prejudice for the child, not the parent. For this ultimate rite of passage, the infant must prepare its lungs, its heart, and its mind for a world it has never experienced. To travel there, it must be shoved out of its warm and protective sanctuary—the amniotic sac—by its mother's uterine muscles, only to be forced once more to descend through a narrow canal that, at its best, resembles a bent cylinder due to the natural curve of the mother's pelvis, making it necessary for its head not only to rotate in its passage through this birth canal, but also to distort and mold itself (because of the softness of the skull) to the contours of the channel on its way out. Its patience, then, is understandable.

White blood cells, which move about like amoebas, crawl about the blood vessels and into the tissues at rates equivalent to moving three times their own length each minute.

For the mother, about to give birth for the first time, the onset of labor—the event leading to the actual expulsion of the fetus out of her womb—is as mysterious as it is unforgiving. It is triggered, most agree, by the fetus when its ten lunar months in the womb have passed. The fetal pituitary gland, prompted by the hypothalamus above it, secretes increasing amounts of the hormones ACTH and oxytocin. ACTH activates the fetal adrenal glands above the kidneys to secrete greater amounts of hormones which, in turn, set off a cascade of events that initiate labor. Meanwhile, oxytocin, in the presence of estrogen (made in the ovaries), stimulates uterine muscle contractions. The young mother feels a dull, low backache and cramps in her belly.

Two weeks previously, this first-time mother, this primigravida in the terminology of obstetrics, might have felt the baby sink lower in her abdomen, and suddenly might have found herself able to breathe more freely—a sure sign of her

Birth was rarely safe or quick. (Above, 1601 A.D.*; right page 1711* A.D.*)*

B
A

impending labor. The phenomenon, known as lightening, was the fetus shifting in the womb to present itself to the birth canal. Like most normal babies, it will leave the womb head first.

Now, two weeks later, she is walking about in the hospital room, anxious and excited. It is already seven in the evening and the powerful rhythmic contractions of her uterine muscles are coming about once every 20 minutes. Each strain of these muscles seems to last an eternity, though, in reality, each contraction lasts only for 15 seconds or so.

A macrophage, a large, bacteria-eating cell of the body, can digest a bacterium in less than a hundredth of a second.

An inspection of the dilation of her cervix, the entrance to her swollen child-filled uterine cavity, reveals that the opening is only one centimeter in diameter, barely enough to admit a fingertip. As labor progresses, it will widen to about 10 centimeters, enough to admit a closed fist or to allow the head of her child to pass through.

As the hours progress, the uterine contractions get stronger, last longer, and come more frequently. The pain increases, but she finds that it is still tolerable. Midnight comes and her cervix has only increased two-and-a-half centimeters—a thumb's width—of dilation.

By 3 A.M., still excited, but tired of the steady, rhythmic pain tearing at her guts, she lies down. A half hour later, the membranous bag surrounding her child ruptures and warm, yellow amniotic fluid gushes from her birth canal. The baby's head presses against the cervix.

The uterine contractions soon come at three to five minute intervals, and each sticks around for some 60 seconds. Confined to bed, she concentrates on her breathing, for when she doesn't, she finds herself panting. The pain in her lower back and abdomen is tremendous.

It is six in the morning. Her cervix is dilated to some nine centimeters. The muscles of her uterus are now flexing every two to three minutes, and each contraction lasts some 80 seconds. She is fatigued and feeling a bit nauseated.

By 8 A.M.—some thirteen hours after her first contractions—the baby wants to come out. The young mother's cervix, now completely dilated and pressed paper thin, allows it to pass. The muscles of the uterus pump the fetus slowly out into the birth canal as if were a wad of solid toothpaste stuck in the tube. Mom helps by "bearing down," using the contractions of her abdominal muscles to force the child out of its residence. It is a behavior that is at first voluntary. Then, her body seems to take over the job by itself. She no longer has to think about performing it. She is wet with perspiration.

Squeezed, molested, and intermittently deprived of oxygen as its umbilical cord gets pressed against the throbbing uterus, the fetus begins to alter its body chemistry. Inside its body, it secretes huge amounts of "stress" hormones, like adrenaline. The hormones speed up its metabolic rate and accelerate the breakdown of stored fat used for fuel. They also clear the fluid from the infant's lungs so that it can breathe

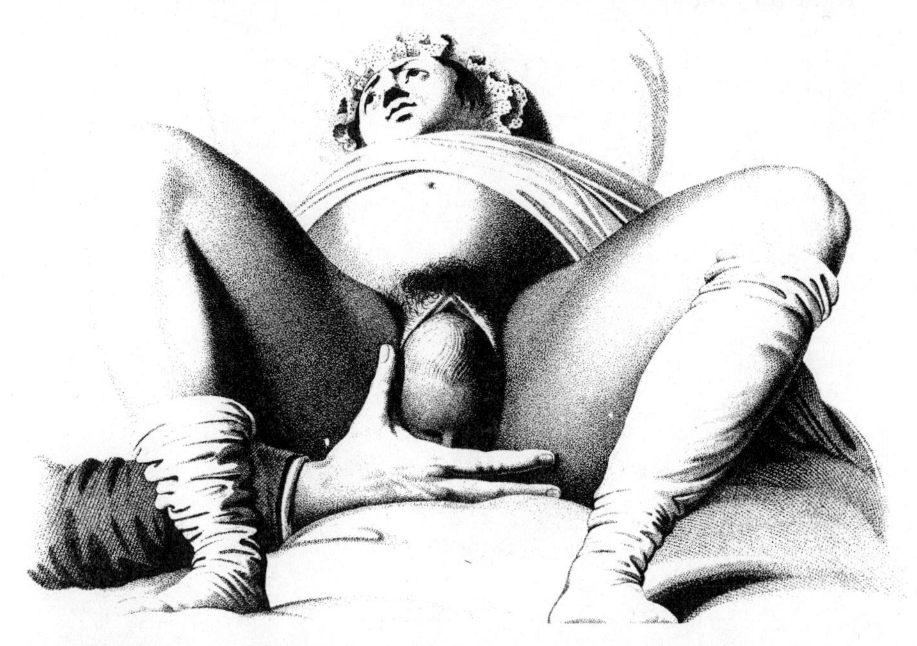

Two pushes and the head is born, another two and the baby is out. (1822 A.D.)

once it gets out. Moreover, as it progresses through the birth canal, its large but soft head molds itself to the inside of the channel.

When the purple crown of the child's head is seen at her vaginal opening an hour and a half later, the woman is all but exhausted. Yet, two more contractions of her uterine muscles, and the child's head is born. Two more, and the body itself is delivered. The child's umbilical cord is clamped and then cut. And for the first time in its life, this perfect baby sucks in the cold, foreign air, and cries.

The journey to the outside has been long and tramatic, and the child soon sleeps. A million years ago, our ancestors probably struggled out of their warm amniotic homes in the same fashion, leaving behind their fluid existence for a world as alien as must have appeared when the first fish walked on land. For our most ancient of ancestors, this monumental step, from liquid to air, offered a new beginning.

For the baby, though, it is morning, a new day. For the mother's, and perhaps for evolution's purpose, his birth was right on time.

SIX

THE BODY IN DAYS

More than fourteen thousand feet up in the Andes mountains of Peru, and this American student, this native of New York City, is freezing, and having a somewhat difficult time breathing in the thin air. He is standing motionless by the blue Ford van that brought him up here along with his six companions, all of whom are looking at the magnificent vista before them. But, the bitterly cold wind is blowing in his face and making it difficult for him to concentrate on the scenery. To stay warm, he pulls the hood of his parka down over his head, but it doesn't help. He is still shivering.

Despite the thin air and the numbing cold, he insists on taking photos of the semi-frozen lake two hundred feet away and of the mountains towering above it. But as he walks briskly toward the water, he suddenly feels a tremendous headache coming on. He gets dizzy and nauseated and feels like he's about to faint. His heart is pounding hard. In his ears his pulse sounds as if huge dumbbells are bouncing off Kettle drums.

The driver of the van yells that it is time to leave. So the student takes a few final snapshots of the beautiful snow-draped mountains around him and runs toward the van. But he covers only about fifteen feet before he stumbles, falls to the ground, and almost faints from the lack of air. Gradually, though, he gets up and begins to walk, but he feels as though he is slogging through hip deep mud. Lifting each of his legs is becoming a real chore, like pulling up tree trunks with his toes. To catch his breath, he takes long, deep breaths of the frigid air. Yet he still feels as if he is suffocating. He is a little nervous.

At the van, two of his six companions—both of them Peruvian women, and obviously wise to the effects of the rarified air (they refuse to leave the van)— are smiling and shaking their heads in disbelief at the behavior of this foolish American.

Seven years ago, I was that naive American tourist. I had traveled up to the "altiplano," as the Peruvians call it, and even higher into the mountains, beyond where even the lichen no longer grow, to sightsee. But, like many visitors up there for the first time, my body was unaccustomed to the thin air.

Strangely, had I stayed up there for a few more days (something I was not about to do), I would have begun to adapt to the lack of oxygen. Then, after thirty or, better yet, ninety days, my body would have almost completely adjusted to the situation. I would have breathed easier and would have been able to walk briskly from lake to van and back again without the least worry. How does the human body adapt itself to such conditions over time?

More than 300 years ago, that question may have been asked by more than a few Spanish conquistadors who, despite their frenzy to destroy the mountain-dwelling Incas of Peru, still feared having to travel these great heights to get them. "There are places inhabited, which are at Peru," one Spanish soldier cautioned in 1604, "where the quality of the ayre cutteth off man's life without feeling." Early in the 1800s, Alexander von Humbolt, the noted cartographer and explorer, described his experiences in the Andes at elevations of 15,000 feet. He observed bleeding of the gums, mouth and eyes in those who suffered some of the effects of oxygen deprivation, known as hypoxia.

One's life is shortened 14 minutes for every cigarette smoked.

In the late 1700s, the cause of hypoxia was unknown, though its malefactor had just been uncovered. In 1774, the Englishman from Yorkshire, Joseph Priestley, discovered oxygen gas. Then, four years later, the Parisian Antoine-Laurent Lavoisier, named Priestley's gas "oxygen" and described its critical function in respiration. But Lavoisier never made the connection between this newly discovered element and the illnesses of his fellow mountain-climbing compatriots.

Enter Paul Bert, a French scientist who, in the late 1800s became interested in the physiology of the body at the upper elevations. During much of the 1870s, Bert conducted experimental studies on humans and animals in the upper atmospheres and was one of the first to discover why the Spanish conquerors and others had so much trouble on their first voyage to these high altitudes.

The real culprit in their inability to tolerate the upper elevations, Bert concluded, was not due to a lack of oxygen per se in the air at these heights. Indeed, the proportion of oxygen is not reduced at high altitudes (it is constant at 21 percent throughout the atmosphere). The real offender is the reduction in the *partial pressure* of oxygen at these heights.

Air is compressible; a tremendous amount of it can be squeezed into a small SCUBA tank, for instance. Because of this capacity, the earth's atmosphere contains a greater number of air molecules at low altitudes than at high altitudes. Therefore, the pressure exerted by the gases of the atmosphere (called the barometric pressure) *decreases* with an *increase* in altitude. By the same token, the partial pressure of oxygen (or

The air gets thinner with increasing altitude.

the pressure exerted by this gas) also decreases with increasing altitude. As the elevation increases, the air gets thinner.

Around 1875, Bert suggested that the diminished oxygen pressure at high altitudes would reduce the amount of oxygen reaching the lungs. This, in turn, would reduce the amount of oxygen available to the tissues of the body. And without enough of the necessary oxygen to feed the brain, for one, it would simply shut down.

To support his point, Bert pointed to a trip that an associate of his, Gaston Tissandier, made in a balloon. Tissandier took two other scientists with him to determine what effects extreme altitude would have on the body. They tragically found out as the balloon rose five miles into the sky. "I soon felt so weak that I could not even turn my head to look at my companions," wrote Tissandier of the frightening experience. "Soon I wanted to seize the oxygen tube, but could not raise my arm. I wanted to cry out, but my tongue was paralyzed. Suddenly, I closed my eyes and fell inert, entirely losing consciousness." Tissandier awoke a few minutes later as the balloon descended to earth, but his two companions died for want of oxygen. So quickly depleted was the oxygen in the blood feeding their brains and the nerves branching from it, that their muscles were paralyzed; they were unable to raise the tubes from the tanks of oxygen to their lips.

Were the body never to compensate for the lack of oxygen at the higher altitudes, no one would be able to live there. Yet, more than 25 million people manage to live and work in the high Andes of South America and the Himalayan ranges of Asia. More than ten million of them live at altitudes above 12,000 feet, and there are mountain dwellers in Peru, who work daily mining copper some three-and-a-half miles above sea level. Moreover, millions of visitors travel to heights of 7,000 feet to camp and to ski and return to their homes at the shore healthy. Obviously, the body is doing *something* quickly to compensate for the drop in oxygen pressure in these individuals.

Understandably, the body's first response to hypoxia is to make sure that the tissues of the body, and especially the brain, get enough oxygen to function properly. The respiratory and cardiovascular systems have evolved a number of mechanisms to do just that. Many of the major blood vessels near the heart are equipped with chemoreceptors, biological gauges of sorts that are extremely sensitive to any drop in oxygen pressure. Alerted by these nerve cells, the muscles of the lungs increase their activity and breathing rate speeds up so that more air will get into the lungs. Moreover, the brain, which has its own chemical meters, orders the heart to pump stronger and faster, so that what oxygen-rich blood there is will get to the body tissues quickly.

But these are temporary measures. Were they the only ones, we would surely fatigue. So the body has evolved other mechanisms to acclimate over time to the thin air. One mechanism involves the red blood cells themselves, the transporters of oxygen to the tissues. For those of us living between sea level and about 6,000 feet, an eyedropper full of blood normally contains some 5 million red blood cells. But, for those living at the higher altitudes, depending on the elevation, the concentration of red blood cells in an eyedropper is much greater. For those living some 14,000 feet above sea level, for example, the number of red cells in a drop of their blood increases to about 7 million.

The process of making more red blood cells to capture and transport what little oxygen there is in the thin air, starts almost immediately. Within the first two hours or so after we begin to experience hypoxia, a hormone called erythropoietin is secreted by the kidney, and to a lesser extent by the liver. The hormone charges the bone marrow to churn out greater numbers of red blood cells. Three to five days later, new red blood cells begin to appear in the bloodstream to reinforce their brothers. But, in about 14 to 20 days, the production of these oxygen-carrying cells levels off as does the secretion of erythropoietin.

The fresh red blood cells come with a special bonus, as well. All red blood cells are packed with the molecule hemoglobin. In fact, red blood cells are little more than sacs stuffed with these molecules. It is hemoglobin that picks up molecules of oxygen from the lungs and then dumps them into the tissues. As one travels to the higher elevations, though, the new red blood cells coming out of their bone marrow homes are made with even more molecules of hemoglobin crammed into them. Moreover, the hemoglobin molecule itself has a property that enables it to take in and unload oxygen more readily when necessary at high altitudes. If a person stays put at these high elevations, these acclimatizing changes take place about 15 to 20 days after one first experiences a little lightness of the head.

All of these changes literally make breathing easier. In a few days, the fast heartbeat slows to rates no different than if one were living by the shore. In about 30 days or so, one no longer has to take deep breaths to feel comfortable.

Blood removed from the body clots in about 6 minutes.

If all of this sounds too good to be true, consider the down side. First, there is the mountain sickness. Mountain sickness, or *soroche* as the natives of the Andes call it, is caused by the lowered partial pressure of oxygen. I had experienced an acute form. The quickening pulse, the labored breathing, the throbbing headache, the weakness in the knees, the nausea—all of which bear a resemblance to seasickness—are temporary. A few moments of rest suffices to banish them.

But for some of the newcomers who come to stay at even higher elevations for an extended period of time, these "mild" symptoms can grow over the hours to bleeding from the lungs and nose, low blood pressure, and ultimately, if the victim remains up there for a few days, to congestive heart failure and death.

Then, there is the loss of appetite, followed by the loss of weight at the highest altitudes. "All members of the wintering party," wrote one investigator during the scientific expedition of the Himalayan mountains in Asia, led by Sir Edmund Hillary, in 1960, "noticed impairment of appetite, particularly for fatty foods." By the end of the first week, Hillary and each of the others had lost some three pounds each. By the end of the expedition, a month later, each had lost about 15 pounds.

Part of the reason for their rapid loss of weight was that the body loses water from the lungs at high altitudes. The air up there is very dry. But weight loss also comes about because, for some unknown reason, taste becomes duller over time at higher altitudes (over 10,000 feet). Sugar, for one, tastes less sweet at high elevations as time progresses. Also, salty, bitter, and sour substances become more and more bland. "As time went on," one mountain explorer recalled, "we developed a marked preference for highly seasoned foods and condiments."

Of all the health effects coming about from living *up there,* though, the most troublesome over time, at least for men, is the loss of sexual function. Indeed, newcomers to the mountains, are much less successful in reproduction than the mountain natives, even after their bodies have gotten used to the thin air. The Spanish conquistadors found this out when they settled in the high Andes and tried to have children. Most of them discovered, to their regret (and to the Incas' pleasure) that they couldn't. They were infertile.

Nowadays, if men travel to elevations above 12,000 feet and remain there for just three days, the level of the male sex hormone, testosterone, in their blood drops by more than 50 percent than what it was when they were at sea-level. When they remain up there for five more days, their sperm count takes a nose dive. Not only do they have fewer than half the number of sperm cells in their testes than they did when they first left the shore, but the numbers of abnormal sperm cells increase. Bizarre sperm cells make their appearance, some possessing two heads or two tails, or no tail.

By the thirteenth day, the number of healthy sperm are reduced by more than a third. After 30 days, there are almost as many abnormal sperm in the testes as there are normal ones. Luckily, if they climb back down to sea level, the sperm count gradually returns to normal, but only after the climbers remain at the lower elevations for more than 25 days.

The loss of sperm cells at high altitudes occurs because the processes involved in making sperm cells are exquisitely sensitive to any decrease in oxygen. Over time, while the rest of the body adjusts to the thinner air, the sperm cells somehow cannot.

The lack of an heir had, no doubt, been a major problem for the Spanish conquistadors and the others who have since tried to build a family in the mountains. But, it isn't always the fault of the man that children are not produced. Apparently, over several days, hypoxia affects women, as well. Women residing at high altitudes, more often than not have erratic menstrual periods or none at all. And in female rats, at least, exposure to rarified air, caused them to bear abnormally small litters of live young, and of those, 20 percent died in the first ten days. That mortality rate is ten times higher than in comparable sea-level groups.

Despite the problems, though, the human body is a remarkably plastic organism. The native Quechua Indians of the Andes and the Sherpa people of Tibet in the Himalayas are evidence of that. Presumably, these people acquire a special physiology for their mountain existence beginning in the womb. Their children are born with a greater number of red blood cells in their body, and with exceptionally large lungs and hearts as compared with children born at lower elevations. The lungs allow them to take in more air with each breath; the heart allows them to pump more blood. With these adaptations, humans have been able to successfully survive at these high altitudes for more than 10,000 years.

Clearly, given time, the human body will adapt to almost any alien environment. Nowhere is that fact more true than in space, where humans have been adapting on and off to weightlessness, or microgravity, for more than twenty-five years. But this "adaptation" has taken its toll on the bodies of the American astronauts and Russian cosmonauts who have stayed up there for any extended period of time. Stories of a few Russian cosmonauts having to be literally carried away on stretchers after the completion of their long missions are not simply rumors.

Most Space Shuttle missions have lasted only seven to ten days in weightless space. But, back in 1974, during one Skylab mission, astronauts remained in space for 84 days, the U.S. record. Still, this duration is a mere pittance compared to the Russians who, in 1984, kept cosmonauts Leonid Kizim, Vladimir Soloviov, and Oled Atkov aloft aboard their *Salyut 7* for two hundred and thirty-seven days—their space endurance record. So how does the body adapt over time to this environment?

The human body is a creature of gravity. Because of its influence, skin sags and leg bones heal vertically, at right angles to the ground. Gravity also tends to force the fluids of the body downward. Luckily, though, two mechanisms work to keep the brain and the rest of the organs of the body bathed in a steady supply of blood. The first, of course, is the heart and the cardiovascular system, which have evolved to work against gravity by squeezing blood under pressure upwards. The second mechanism involves the normal tissue pressure in the legs themselves, which counter the tendency of gravity to pull the body's fluids downward.

In the skin, blood flow regulates body temperature. When it is cool outside, blood flow in the skin is about a half a quart a minute. When it is warm outside, blood flow increases to cool the body to roughly three quarts of blood per minute.

But immediately after the body becomes weightless, these same pressures from the tissues of the legs shifts the blood and other tissue fluids headward with the result that the faces of astronauts puff up. They also experience a stuffy nose. Gradually, though, these fluids are passed off in the urine and in the water vapors from the lungs, normally within four to five days. Eventually, the stuffy, bloated feeling wears off.

To compensate for the sudden shift of fluids in the body, and for the stress from being in this novel situation, the brain triggers the secretion of a number of hormones from the body's endocrine glands. Hormones like adrenalin, produced by the adrenal glands, and ACTH, produced by the pituitary gland, surge out into the bloodstream. For a while the heart beats rapidly. But, gradually, over the course of a day or so, the hormonal output adapts to the new situation. The brain accommodates to the microgravity.

For the first two days, visitors to this weightless environment are slightly disoriented, because in space the body has

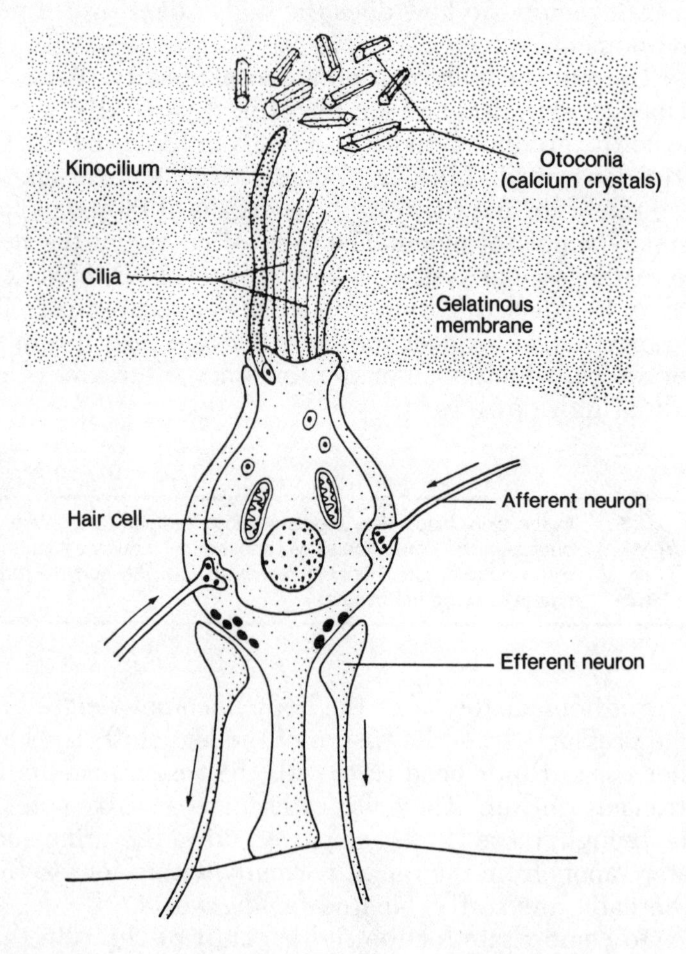

It takes days for the brain to adjust to the signals from the vestibular system in the inner ear.

no way of telling which way is up and which is down. An astronaut climbing a ladder may have the strange feeling that he is descending headfirst. Shutting his eyes, he may feel as though he is falling.

The unsettling feeling of disorientation is caused by a glitch in the vestibular system. The vestibular system, responsible for telling the body that it is tilting or swaying, requires gravity to do its job. Within the inner ear sits the saccule and the utricle. Each contains a tiny patch of jelly embedded with tiny grains of calcium called otoconia. Beneath these, hair cells implant their "hairs" into this gelatinous substance. Normally, changes in the position of the head cause the force of gravity to distort the position of the jelly. Like a glob of Jello sliding across blades of grass, this distorts the hairs of the hair cells. The contact then generates nervous signals that, along with other cues, produce an overall sense of orientation.

Taste buds are among the earliest sense organs to appear in the fetus. By the third trimester of pregnancy, fetal taste buds are responsive to chemicals in the amniotic fluid.

But when gravity is taken away, the gravity-sensitive organs of the ear no longer send the correct signals. The eyes tell the body that it is moving up, but the vestibular system does not confirm it. So, for the first 2 to 4 days, the body is somewhat disorientated. In severe cases, as has happened with roughly half of all of the space travelers, space sickness, a kind of seasickness, sets in. Those affected feel dizzy and nauseated for the first couple of days. Gradually, though, the feeling inexplicably disappears. Somehow, the brain learns to rely on the new signals coming from the vestibular system.

But by far the most disturbing consequence of prolonged weightlessness is the effect it has on the bones. Bone, which is primarily strengthened by calcium, is maintained by stress and pressure. As stress lessens, bone dissolves and calcium washes out and is excreted in the urine. The bones eventually become brittle. In space, since nothing presses down on the bones of the body, they demineralize. In one study, researchers found that the amount of calcium in the urine of *Skylab* astronauts increased by 100 percent, reflecting an overall bone loss of nearly a half a percent a month. Even vigorous exercise could not alleviate the metamorphosis.

Still, as insidious as these changes seem to be, the body is simply adjusting to a new environment. The time, a matter of days, is required to adapt a physiology well-suited for one situation, to a new one. Whether it be on a mountain top or inside a space ship, the body, over time, will conform.

Someday, perhaps sooner than we think, some of us will take up residence on the Moon or on Mars. Knowing how plastic and adaptable the body is to its new surroundings, we can only wonder what nature of man they will become.

SEVEN

THE BODY IN MONTHS

On the seventeenth of December 1982, in a southern California hospital, a child was born. The boy was one of a number of children born that day. Yet, for both the parents and the hospital staff, the event was extraordinary. The boy's birth was three months premature.

Lying in the enclosed warmth of the ward's incubator, the infant looked brittle and frail, and so tiny (he was as small as his mother's hand), his parents were scared to hold him. The hospital scale put his weight at just one pound, seven ounces. Machines kept track of his temperature and heartbeat. Plastic tubes fed nutrients into his bloodstream. Unable to breathe and lacking the muscle tone of normal infants, the child required a respirator to forcefully pump air into his immature lungs.

The infant, Dustin Kemp, survived his early entrance into the world, eventually acquiring the plump, cherub health common to infants carried full term. He was the exception, though. Few children born so young have survived as well, if at all. But Dustin was the exception in another sense. His early birthday points to what is peculiarly unique about the biology carried out within the human womb. It follows a specific developmental agenda that is only fully completed after an average of nine lunar months.

Among the primates, humans spend the longest period of time within the womb. Lemurs, cousins to the earliest of primate ancestors, have a period of gestation lasting just four months. Macaques and squirrel monkeys spend no more than six months within the womb. Gibbons, members of the great apes, are born after only seven months of prenatal life. The chimpanzee, one of man's closest primate relatives, has a period of gestation just a few weeks shorter than ours. So, why is ours so long?

The brain produces cerebrospinal fluid at the rate of a half quart per day.

Bigger Brains

Evolution has endowed humans with large brains and, in particular, a sizable cerebral cortex. Our complex nervous system is one of the few characteristics that makes *Homo sapiens* so different from the other primates. Only a young *human* can ask questions about itself (Where did I come from? or Why do I look like the mailman?). The most talented five-year-old chimpanzee, taught the sign language of the deaf, can only communicate something on the order of: "Me eat, drink more." So, if there is a pattern here, it is that humans possess big brains and bigger brains call for a longer gestation period.

The problem with this line of thinking, however, is that it doesn't completely fit the evidence. At birth, human brains are not much larger than the brains of most baby apes. In fact, the average brain size of a newborn child, is only slightly larger than that of a newborn chimpanzee. The argument becomes even more muddled when one looks at the bodies of human babies. They are, in all respects, disproportionately bigger at birth than their ape cousins. So, it looks as though the point should be made this way: Bigger, more complex, bodies must provide bigger brains. Both bigger brains and bigger bodies must require a longer period of gestation. Maybe.

For humans, larger fetal torsos were a nice evolutionary touch. The excess of body fat they harbor, equips them with a ready supply of energy immediately after birth. But large bodies would have been nearly impossible to achieve without a similar trend towards an increasingly larger birth space in the mother, and a move toward single births, a phenomenon that we share with many of the great apes.

For *Homo sapiens,* though, evolution dealt a mixed hand. We are bipedal creatures. For our ancestors to be able to walk upright, though, the pelvis had to become narrower. But, as the hips shrunk through the millennia, the birth canal grew more cramped at the same time the brain and skull were becoming progressively larger. The result is that there is now a hair's difference between the diameter of the human birth canal and a newborn's head. Successful delivery is only possible by softening the ligaments that hold the bones of the pelvis together shortly before and during birth. But despite the evolution of a pliable pelvis, labor for the human mother can still be excruciatingly painful and delivery damned near perilous.

Watching squirrel monkeys give birth—primates that also have small pelvic openings—is like watching evolution gone overboard. Late at night and about four hours before the onset of labor, a pregnant female will move away from the others in her cage and lean against the wall in silence. When the pulsating contractions of labor finally arrive, the mother freezes in place, occasionally arching her back. She opens and closes her mouth, uttering no sounds. For as long as two hours, the steady contractions of her uterus persist, squeezing the infant slowly out from her swollen belly through the birth canal. Unlike human births, though, the face of squirrel monkeys appears first. Some mothers grasp it and begin yanking both head and body free of the birth canal. Sometimes, an infant whose hands have escaped the birth canal, grasps hold of its mother's belly fur and begins to pull the rest of its body out. After a few minutes

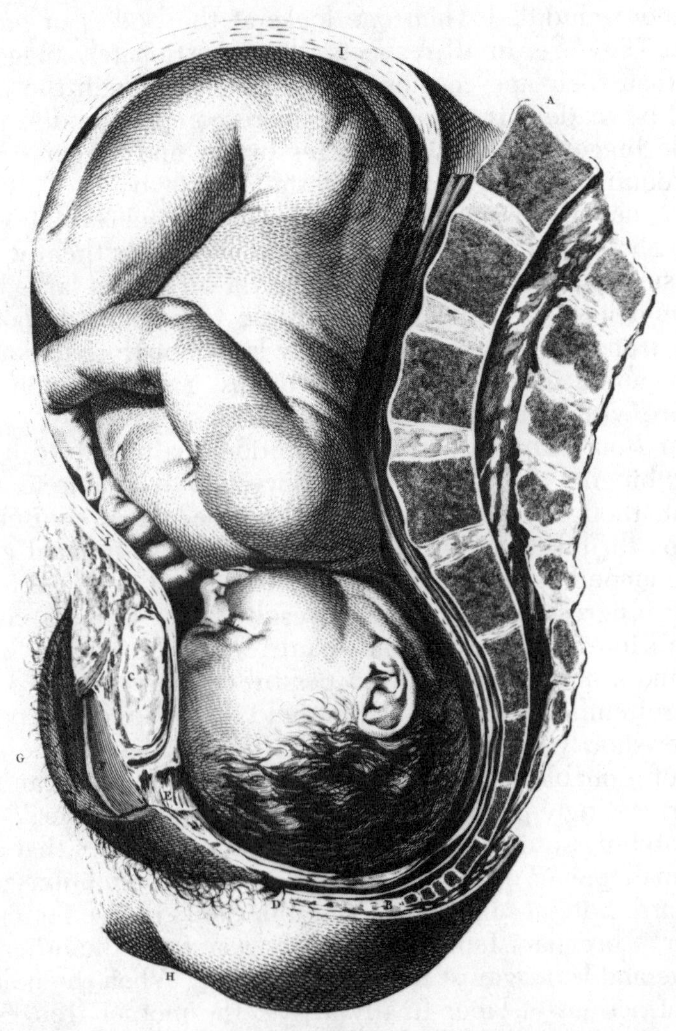

There is virtually a hair's difference between the fetus's head and the diameter of the birth canal (drawing 1754 A.D.).

of this struggle, it is over. The infant emerges, its fur matted and damp, and begins searching for the warm milk from a mother who is obviously exhausted from the whole ordeal. The infants that make it this far, though, can be considered the lucky ones. As a rule, more than half of the fetuses die as a result of cephalopelvic disproportion. With skulls larger than their mother's birth canal, they never make it out of the womb. The mothers and babies die soon afterwards.

Cephalopelvic disproportion occurs to a far less degree in the great apes. Newborn gorillas, for instance, are exceptionally tiny and their heads are smaller than the pelvic inlets of their mothers. Delivery is quick and easy and labor, a traumatic experience in monkeys, is often hardly even detectable.

So if we've found ourselves out on some precarious evolutionary limb far from the great apes, it is not without reason. We have balanced the need to walk upright with the advantages of bearing larger offspring. But we are left with the same nagging dilemma. If we are born with a skull no larger than the small pelvic canal of our bipedal mothers, why do we need the longer gestation time?

Womb Life

Much of the answer lies within the warm, liquid confines of the human womb. Here, a program of development is carried out that, by all accounts, mimics the evolution of animal life itself, something German embryologist Karl von Baer first discovered in the nineteenth century when he deliberately left off the labels of embryonic specimens in various stages of mammalian development. "I am quite unable to say to what class they belong," he wrote in his *History of the Development of Animals* in 1828. "They may be lizards, or small birds, or very young mammalia, so complete is the similarity in the mode of formation of the head and trunk in these animals." Von Baer was also the first person to identify the human ovum and the first to follow a fertilized egg, or zygote, through its development. In its journey, von Baer saw one form melt gently and inexplicably into another; each step always following the last; no step missed, no step skipped. Only many years

Monkeys, unlike many apes, give birth to their offspring with difficulty.

later would researchers discover the molecules of heredity that direct an embryo's voyage or identify some of the migrations and divisions of cells that sculpture it.

By now, much of the developmental program of human prenatal life has been mapped out in excruciating detail. What's been found is that our life in the womb is much more active and advanced than we once believed.

An infant goes through three major stages before it leaves the womb. During the first stage, the zygote divides and divides and divides until it is a mass of cells, a mulberry, no larger than the original egg. Three days later, this ball of cells hollows out, then puckers up. In the process, it forms a kind of deflated and punched in soccer ball with its cavity sealed. By the end of the first week, this strangely formed cluster settles onto the wall of the uterus, now thick and hormonally prepared for its arrival. The cluster then attaches itself, swiftly burrowing deep into the uterine wall, dissolving away the few cells lying in its path. By the twelfth day after fertilization, it is snugly dug into the uterine lining.

After this strange-looking mass is comfortably settled, the second stage in the life of this tiny "womb-ling" begins, this time as an embryo. For the rest of the first month until the end of the second, cells begin to divide and sort themselves out into tissues and organs according to the instructions of their DNA. Those cells destined to become members of the kidney, do so. Others, which look no different from the cells around them, suddenly organize themselves into a miniature spleen or liver. Their neighbors assemble into a tiny heart and begin to beat in unison. Lips form, elbows develop, arms bud, fingers evolve—in an infant no larger than a dime.

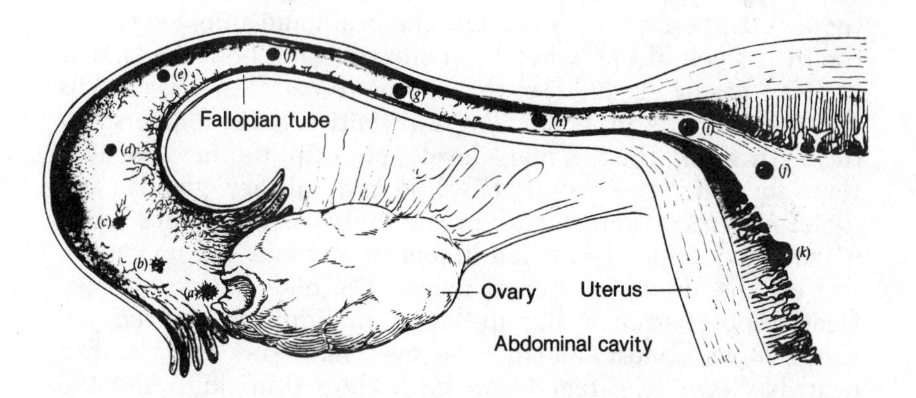

The egg cell leaves the ovary and is fertilized in the fallopian tube. Dividing, the mulberry eventually burrows into the uterine wall. (From Grobstein/External Human Fertilization.

If the infant looks finished, though teeny, it is far from it. In the human embryo, the cells of the body parts—arms, legs, liver, lips—may know their identities by the end of the second month, but the cells forming the brain are as yet uncommitted. In fact, even though organs like the eyes and ears are fairly well completed by the second month of life, the brain is still not as yet connected to them. In a sense, the phones are installed, but the lines to them are not. So, the brain lags behind the rest of the body in development. Why? Apparently, the human brain needs to remain plastic, or (excuse the pun) open-minded so as to constantly adjust to the changing environment around it. To do this, it is programmed to go through an exceedingly large number of complex changes over a long and extended period of time.

Certainly, unsuspecting mothers are not carrying around zombie children in their womb for two months. In fact, as far back as the third week of life, when the human embryo is nothing more than a two-layered disk smaller than the eye on a Lincoln penny, the nervous system is recognizable, though it is but a thin sliver of cells lying on the embryo's surface. In a few days, this so-called neural plate stretches somewhat to form a shallow gully on top of the embryo known as the neural groove. By the end of the first month of life, when the heart is just beginning to beat, the folds on either side of this gully close over and fuse together forming a kind of tunnel. Many of these cells then go on to become the brain and spinal cord. By the fifth week of life, when the embryo (now about the size of a raisin) has hands and feet that are mere paddles, and nostrils that are but pits in its surface, a primitive brain appears. By the sixth week, the nose has developed a tip, the face has tiny flaps for ears, the mouth has lips and salivary glands, and something marvelous occurs in the embryo. It flinches when it's touched. It possesses a rudimentary nervous system.

By the time the second month has ended, the embryo, floating now astronaut-like in the warm, liquid sack of the placenta, looks almost human. The eyes have tiny eyelids. The head has jaws, the feet toes, the body a thin skin. And the cerebrum of the brain—the future site of judgment, intellect, and reasoning—has split into two dinky hemispheres, each about the size of a sesame seed.

All that is left is for the tissues and organs, established during the embryonic period, to grow and mature. That is exactly what happens. For the remaining time within the womb, the infant, now called a fetus, grows in size and complexity.

During the third month of life, teeth form under the gums. Fingernails appear. Kidneys release urine into the amniotic fluid. The nostrils, once no more than nasal cups, now fuse with the throat. The loops of intestine, which had their home within the umbilical cord, slip into the fetal abdomen.

Fingernails grow considerably more slowly in arctic climates than in temperate ones. Blood flows through the skin to a lesser degree in the cold than it does in warm weather. Therefore, cold skin receives less nutrients, and fingernail growth suffers for it.

By the end of the fourth month, the fetus can swallow and, in fact, gulps down amniotic fluid by the teaspoon, actually using it as a nutritive supplement in its diet. It "breathes" weakly.

By the waning days of the fifth month, its legs are moving around enough to be felt by its mother. Its heartbeat can be heard with a stethoscope. It grows coarse hair on its head.

By the sixth month, the skin is covered by fine downy growth of lanugo hair and by a greasy, cheesy substance made of the secretions of sebaceous, or oil glands of the skin, known as *vernix caseosa.* It is about the size of a loaf of bread now and weighs almost two pounds.

After that, events accelerate. The eyes, which had been closed since the third month, suddenly open. If it is a boy, fetal testes descend into a scrotal sac. If it is a girl, fetal eggs multiply within fetal ovaries (to someday make future fetuses). Lips suck an invisible nipple. Hands grasp unseen objects. Lanugo hair disappears. The tongue moves. Gut enzymes activate. It gets bigger. Lungs develop. Glands enlarge. Skin darkens. Hormones flow. Then, birth!

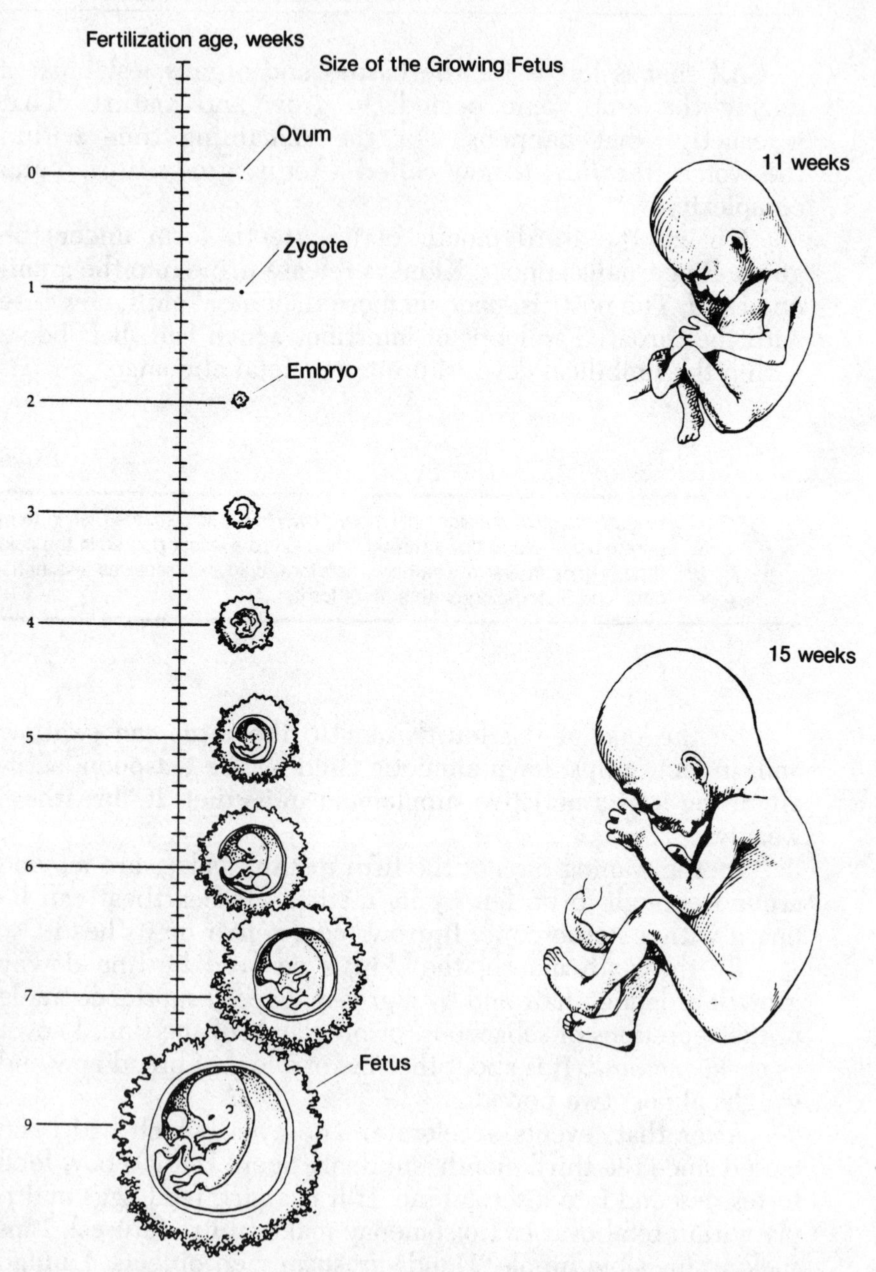

Growth in the womb is a slow, step-wise progression. Source: Patten, *External Human Fertilization,* copyright © 1968 by McGraw-Hill Book Company. Used with permission.

That is the picture of development from the outside looking in. But while these extraordinary events are occurring, there is something fantastic going on within the hidden confines of the fetal brain. It is sorting itself out, organizing, seeking an identity.

At the beginning of the fetal period, the cerebral cortex is in a primitive state. For a while, the thousands of small, spherical cells of the cerebral cortex do nothing more than multiply. But by the end of the third month, though, these so-called neuroblasts migrate out in waves, forming layers of cell bodies as they make their way toward the surface. Meanwhile, deep within the recesses of the brain, close to the developing spinal cord, thin, stemlike axons from nerve cells more mature than those in the cerebral cortex, advance like tropical vines into the dividing mass of the cortex.

By the fourth and fifth months, cells throughout the brain are multiplying at the incredible rate of nearly 300,000 new cells per minute (by birth, the brain will contain all of the nerve cells it will ever need, about one trillion of them), only to wind down—slightly—by the end of the twentieth week of life. Then, the encroachment begins. From each region of the brain, nerve cells send their spindly axon arms to invade another. Throughout, cells greet. Many synapses form. The roots of nerves infest the brain. A network is built. The body moves.

By seven months of age, the brain is entirely active. Scientists can actually record brain waves in the fetus that are strikingly similar to the brain waves of newborn infants. From the brain's electrical activity, they can record the fetus dozing in REM sleep. Ultrasound pictures, made by sending high-frequency sound waves through the womb and bouncing them off the fetus, show an infant sucking its thumb, hiccuping, smiling, even stretching its limbs. It feels its surroundings, touching and actually avoiding getting wrapped around its umbilical cord. Shine a bright light on its mother's tummy, the fetus will shut its eyes. It may even raise its hands to its face.

What excites researchers most, though, is that a fetus can hear in the womb—its mother's heartbeat, her grumbling stomach, her rhythmic breathing, her voice—and that it remembers what it hears. In 1980, in a laboratory in the University of

North Carolina, psychologists Anthony DeCasper and Melanie Spence tried an ingenious experiment. They asked 16 pregnant women to read the children's book *The Cat in the Hat* by Dr. Seuss out loud, twice a day, to their swollen bellies. They asked them to do this every day for the two months remaining before they were scheduled to give birth.

When the babies were born, the two researchers connected the infants up to a system where they could suck on a plastic nipple attached to a tape recorder. If the baby sucked on the nipple one way, it would be rewarded by hearing its own mother's voice reading the Dr. Seuss tale. But if it sucked another way, it would hear its mother reading a popular postnatal favorite called *The King, the Mice, and the Cheese.* Each had a poetic meter quite different from the other. Remarkably, the children sucked on the nipple to hear their mother read the story they had heard first in the womb. Either Dr. Seuss is the darling of the embryonic set, or infants are able to remember what they hear from within the amniotic sac.

After brushing the teeth, dental plaque begins to develop within six hours, and more rapidly at night.

So we are left with an evolutionary riddle. We possess a fetal brain more advanced than we had imagined, yet one that is still wholly underdeveloped by the end of gestation. Most of the cellular supporting structures of the brain have yet to be established. Furthermore, at birth the brain is, on average, no larger than a chimpanzee's, but it contains many thousands of nerves and nerve connections more than its primate cousin's.

We are, it seems, an evolutionary compromise. Harvard University paleontologist Stephen Gould, has suggested that the increase in brain size that occurred during human evolution was the result of a prolongation of rapid prenatal growth. And yet, fully three-quarters of brain growth in humans is postponed until *after* we are born.

The reason for this happens to be purely obstetrical. In order to pass through the pelvic canal, a human neonate must be born at a more immature stage. In other words, the brain has to remain small so that the fetus can get out of the womb. The fossil record backs this up. The brains of newborns in comparison to those of adult brains, have actually decreased in size from *Australopithecus,* a three-million-year-old ancestor of ours, to *Homo sapiens.* Therefore, our large, complex brains are the result of an increase in *postnatal* brain growth.

The tight packaging of the human brain before birth, then, is a neat evolutionary trick. The brain remains small enough for the child to get through the birth canal, yet it is still full of potential. The limited growth of the human fetal brain must be designed, therefore, to provide a basic network for the more complex maturational changes and excessive growth of the brain to come. The timing of our births is probably triggered when this essential network is complete. Only then are we prepared to face the external world. In sheep, for example, the time of birth is under the control of the fetal brain.

Of the primates, we alone possess both the largest brain and the longest lifespan. It may have been that, in the evolution to our present form, these two attributes went hand-in-hand. Our extended lifespan provided a longer period of time for brain growth and development to occur within the womb and for it to spill over into childhood. As a result of this drawn-out period of brain growth, adults of our species have larger and more complex brains. And, no doubt, the larger brains provided humans with the greater intelligence to successfully care for, nurture, and protect the next generation of our slowly-developing species.

Seasons

The advantages in prolonging the duration of prenatal life were so important to the evolution of humans, that it seems only natural to suppose that there might have been a similar advantage in the seasonality of birth as well. Dustin Kemp, for one, was born in winter. As it happens, so were many other people.

There is an obvious rhythmicity to the earth. The tilt of its axis and, along with it, its annual revolution around the sun, produces the seasons. The days get longer, then gradually shorter. So it would be foolish to think that the plants and animals living on a planet that goes through such changes, would not themselves be affected by them.

For most birds, the height of the breeding season is spring; for deer the mating season is the fall. Brook trout spawn in the fall; bass in the late spring and summer. That increasing day length increases the development of the sexual organs and the migratory behavior of birds in spring was demonstrated some

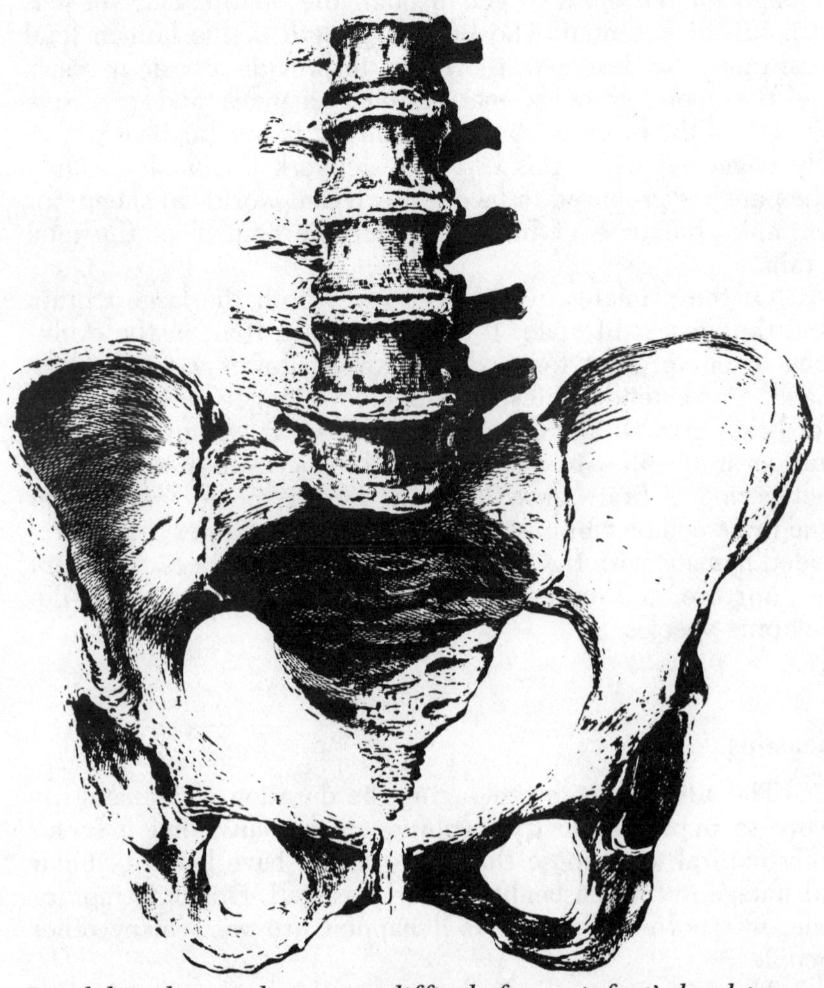

Bipedalism has made it more difficult for an infant's head to pass through the opening in the female pelvis (drawing 1754 A.D.).

fifty years ago when biologist W.R. Rowan forced finches to breed out of season by artificially increasing the length of time he kept the lights on in his laboratory. After the breeding season, the gonads of birds spontaneously regress to an immature state.

Gonorrhea has an incubation period in the body of 2 to 9 days.

The behaviors of animals (and, of course, plants) are not dependent on the length of day or night, as such. Rather, what seems to be involved is a circadian rhythm of sensitivity to light as the inducing or inhibiting agent. Most researchers now believe that animals measure time internally starting at the beginning of either the break of day or the beginning of the night. A brief period of time is needed, however, to trigger a sensitive stage that responds specifically to light. The breeding period of the ferret, for example, can be started prematurely by exposing the animal to 12 hours of artificial light each day for a month if only one hour of light is first given from midnight to one in the morning.

The seasonal behavior of most animals is taken for granted. The evidence that human physiology follows the seasons, though, is far more enigmatic. Part of the problem lies in the fact that most humans live in the city, an environment ruled more by the clock than by the sun. Moreover, culture, religion, national practices, and economy, all play a role in human behavior. The trick for biologists, then, is to separate the influences of the artificial manmade environment from raw human physiology. When they do—by using large samples from all over the world— some interesting findings keep cropping up.

On November 9, 1965, New York city suffered its most devastating black-out. The lights were out everywhere— television, radio, refrigerators, clocks—nothing worked. The residents were left with nothing better to do that balmy, moonlit night than to crawl into their beds and snuggle under the covers.

About a year later, someone was looking at the records of hospital births in the city and discovered something pretty exciting. The data indicated that there was a surge in the number of births in the city exactly nine months after the black-out. The implication was obvious. People had been sleeping around in the City that never sleeps. When the press got wind of the story, it made headlines throughout the world. It's been part of the folklore ever since.

The funny thing was that while the numbers were correct, the interpretation wasn't. When scientists began to look over the several years of data, they discovered that the same surge in births occurs every year in the City and during the same month. There was no extraordinary call of the wild that day in November. New Yorkers have lots of sex during that time of the season every year.

Isolated incidents are one thing, but there is evidence that a number of aspects of human reproduction vary seasonally over the year. The numbers of abnormal or nonliving sperm in fertile men, for example, are highest during the winter. The birth weights of neonates delivered in the summer and autumn tend to be slightly greater on the average than those delivered during other seasons of the year, as well.

There is a universally observed seasonal or circannual rhythm in the number of human conceptions each year. Mapped out by P. H. Jongbloet at the Free University in The Netherlands, the highest conception rates peak in the spring with a lower peak occurring in the fall. Jongbloet's observations support the age-old belief in "spring fever" seen especially in college students. Why that is, isn't really known. But, if we have adopted the seasonal rhythms of lower animals, our species may be affected by the seasonal change in day length.

In many male mammals, sexual activity coincides with a rise in the activity of the testes, which are themselves affected by day length. Day length influences the activity of the pineal gland. The pineal gland, found in the brain, secretes the hormone melatonin during periods of darkness and is thus sensitive to light or the lack of it. According to one theory, the seasonal activity of the pineal gland influences the activity of the testes. The increased size of the testes in spring, for example, would cause a subsequent increase in the levels of the sex hormone testosterone found in the blood. That it happens in species from bullfrogs to baboons has been known for years, but does it occur in humans?

In 1978, Alain Reinberg and his colleague Michel Lagoguey set out to study just that. For more than a year, Reinberg and Lagoguey measured the sexual activity of five male medical students. The students had volunteered to document their sexual activity and to have their blood and urine samples tested each month for signs of testosterone. Each day the budding medical doctors checked off how many times and at what hour of the day and month they had either masturbated or fooled around.

When the study had ended, Reinberg and Lagoguey found that there was not only a seasonal rhythm in the body levels of testosterone, but a daily rhythm, as well. They discovered that the highest levels of testosterone occurred in the early morning (around 8:00 A.M.) in the spring, but in the early afternoon (around 2:00 P.M.) in the late fall. Moreover, when they looked at the levels of testosterone over the entire year, they found that the levels of this sex hormone were highest in October and lowest around May.

At the time of birth, girls have about 400,000 egg cells in both ovaries. Of these, only about 480 may ovulate during her entire reproductive life and of these, only five percent will be fertilized by sperm.

As it happened, the levels of sexual activity matched the levels of testosterone in the body. The medical students reported that they had masturbated or had sex most frequently in the fall but, surprisingly, fooled around less frequently (though they *did* fool around a lot) in the spring. While Reinberg and Lagoguey's study throws a damper on the phenomenon of spring fever, they still concluded that human sexual activity is influenced, though perhaps not in a major way, by the effects of a seasonal change in day length on the activity of the gonads.

If there is a seasonal rhythm in human conceptions, one would expect to see a seasonal rhythm in human births, as well. Jongbloet, for one, noted that the highest rates of birth occur around the winter with a lower peak occurring in the summer. If true, the cycles may have had some adaptive value. Children who were born in the winter months, for example, would spend their most critical childhood months during the spring and summer when there is lots of food available for them to eat. The "peak" babies would then be stronger and more successful than their "trough" brothers and sisters.

If success were measured by office, then one would expect, too, that most of the Presidents of the United States would have been born during the "peak" months. Not true. A glance at the seasonal placement of their births, shows that nearly as many Presidents were born during the seasonal "peaks" as were born in the "troughs." While the "peak" babies included Washington, Madison, Lincoln, and F.D.R., "trough" babies included Jefferson, Monroe, Truman, Kennedy, and L.B.J.

Presidents do not, as a group, represent a cross-section of the human race to be sure, but they do indicate how muddled the studies of modern human seasonal variations have become. On stronger footing, however, are the rhythms associated with children born with congenital malformations. Oddly enough, these seasonal variations are also present in the southern hemisphere. As expected, they are shifted by six months.

The body loses water through the skin (from simple diffusion) at the rate of a half quart per day. The body loses water from the lungs by the same amount, in the breath.

For years, physicians have noticed a seasonal trend in the births of malformed children. Children born with spina bifida, a problem where the spinal cord is exposed, and eye cataracts, most commonly have birthdays in the winter. Children born without limbs or with deformed ones, or with congenital heart problems, or even with abnormal intestines are, inexplicably, most likely to be born in the summer.

Of the most often cited congenital malformations, the one most investigated has been that of congenital hip dislocation or CDH. Strangely, data collected throughout the northern hemisphere consistently show the same thing; that CDH is most likely to appear in infants born during the autumn. Just as oddly, data collected from Australia, in the southern hemisphere, show that CDH occurs there exactly six months later, during their autumn.

Why congenital deformities should occur more often during one season than another is one of the mysteries of life. It may, as some researchers have speculated, be related to the seasonal occurrences of infectious diseases in children, or even due to something as benign as the annual cycles in the weather. But those theories have lost favor in recent years. A more likely explanation is that the malformations come about because of seasonal changes in the physiology within the womb, including seasonal hormonal changes in the mother. The theory suggests that seasonal differences in the physiology of the pregnant mother, affect the development of the fetus. Asked for proof that this might be the answer, researchers point to the work of Harold Kalter at the Children's Hospital Research Foundation in Ohio.

In the late 1950s, Kalter was studying the effects of the steroid hormone cortisone on the offspring of pregnant mice. One of the effects cortisone has is the tendency to induce cleft palate in baby mice treated with the steroid before birth. But for some reason he did not understand, Kalter kept getting conflicting results each time he experimented. Sometime the cortisone induced the malformity; sometimes it didn't. Perusing the data, it suddenly occurred to him that the results formed a seasonal pattern. He decided to confirm his hunch.

Kalter gave each member of a group of pregnant mice exactly the same dose of cortisone at different times of the year. When he looked at his results, he was amazed. His discovery corroborated his suspicion. The effects of cortisone treatment on his mice offspring were indeed seasonal. Far more mice were born with cleft palate in those groups given the steroid between November and April than when he treated their mothers with cortisone between May and October. The conclusion: Either the mother, the developing infant, or both are more susceptible to the body's hormonal environment during one part of the year than another. Recently, other researchers have found similar seasonal susceptibilities to other chemical substances.

If these dramatic seasonal differences in congenital deformities tell us anything at all, it is that modern humans may still possess the remnants of the reproductive biology of our distant ancestors. While nearly all women of reproductive age menstruate, for instance, some woman have been found to release eggs only in the spring or late autumn, and hardly ever in the winter and summer. This may mean that humans have ovulatory and non-ovulatory seasons in which woman produce more eggs during one part of the year, less during another. The seasonal congenital malformations may be the result of a seasonal production of overripe eggs.

Watching the seasonal breeding of our primate relatives offers some explanation for our predicament. Rhesus monkeys breed in the winter and then ignore each other in the spring and early summer, a behavior which matches the biology of their bodies. More than half of the eggs taken from the ovaries of a female rhesus monkey and placed in a culture dish ripen during the breeding season, while only a small percentage do so during the nonbreeding season.

After a person has drunk a glass of water, as long as a half an hour to one hour is required for all of the water to be absorbed into the body from the gut.

Like the monkeys, our ancestors may have had an adaptive reason for seasonal sex and birth. Living under the harsh conditions of the African savannah, their seasonal behavior could have been valuable not only in finding enough food to eat, but in keeping the family together.

A few million years later, we find ourselves able to breed throughout the year—successfully. But our sexual behavior is not immune from the seasons, and our biology definitely is not. We still follow the gentle rhythms of the months. Hormone secretions rise and fall throughout the year. Our gonads still follow the sun.

A Season for Illness

Were sex and birthdays the only evidence for the seasonal variation in human biology, we would surely be unique among the animals. It isn't. We aren't. Allergies, infections, heart troubles, and pain, are more common during one season of the year than another.

More than twenty years ago, European scientists took note of the seasonal changes of asthma sufferers in the Netherlands over an eleven-year span. Most attacks, they found, occurred in the late summer and into the fall. Few attacks took place in the spring. Many researchers were skeptical. They assumed that the data must have reflected a seasonal variation in the amount of dust, pollen, and otherwise nasty particles which many asthma sufferers are sensitive to. Then, in 1978, Alain Reinberg, Michel Lagoguey, and a number of other scientists looking into the seasonal attacks, found that they were not so much triggered by the amounts of antigens in the environment, but were, in many instances, a natural consequence of the rhythmic cycles in the sensitivities and activities of the molecules of the body's immune and hormonal systems.

Some people, for instance, are hypersensitive to house mites, microscopic bugs that infest by the hundreds of thousands everything from carpets to the stuffing in mattresses. Their human victims sneeze, get itchy eyes, and have trouble breathing. Mostly, their attacks come in the winter, when the victims naturally spend more time indoors and are thus exposed to the mites. But, strangely enough, very few of these sufferers exhibit allergic symptoms in the summer, a time of year when there are far more mites inside carpets and beds than during any other time of the year.

University of Texas environmental scientist, Michael Smolensky, for one, argues that the allergic response to these mites and to other allergens may also be the result of some seasonal rhythm in the body's immune response. Indeed, when Alain Reinberg analyzed the blood samples gathered every other month from nine young Parisian men and women, he found a seasonal rhythm in their white blood cell counts. The highest counts occurred in the fall and winter months. White blood cells are the major constituents of the immune system.

That the immune system follows the seasons is a fairly new observation. If truly universal, it would account for the seasonal occurrences of infectious diseases seen in humans, an observation noted by the Greek scholar, Hippocrates in 400 B.C. Today, the Center for Disease Control in Atlanta, Georgia keeps tabs of the numbers and trends in diseases as trivial as the cold or as potentially deadly as polio. Summarized in its *Morbidity and Mortality Weekly Reports* and confirmed elsewhere, these contagious diseases are mapped out in detail and studied at length. What has been found is that there is a seasonal rhythmicity in the incidence of these diseases.

Scientists and public health specialists alike have found that contagious diseases such as the flu, cold, and pneumonia, are more likely to occur during the winter months than during the summer. On the other hand, childhood diseases such as chicken pox, mumps, rubella or German measles, and rubeola (measles) are more likely to occur during the spring and summer. Whether or not these diseases are entirely due to the seasonal changes in the body's biochemistry isn't known, but interestingly, the levels of antibodies for rubella, still rise and fall with the seasons in individuals who no longer have the disease.

A knee jerk takes about four tenths of a second to complete itself.

Like the contagious diseases of children, adults have their own uniquely grown-up infections. Mature and often sexually active, they suffer the consequences of their behavior. Among young adults, venereal diseases such as syphilis and gonorrhea follow the seasons.

Gonorrhea, familiar to the Chinese more than 5,000 years ago, and given its name by the Greek physician Galen in 200 A.D., is a disease initiated by microorganisms that like to swim around in the mucous membranes of the urogenital tract. Syphilis, a disease unknown in Europe before the return of Christopher Columbus from the new world (some say his crew brought it back with them from the Americas), is caused by another microorganism that travels into the bloodstream, producing ugly lesions on the skin, and sometimes death. Predictably, the two diseases are most prevalent in early August and least prevalent in March; predictably because their rhythms mirror the circannual rhythms of human sexual activity. Still, no one as yet really knows whether the seasonal patterns of these diseases are due to circannual changes in the virulence of these sexual pests, or to the rhythm of the immune system which protects the body from them.

It takes approximately 60 seconds for blood to circulate throughout the body at rest, faster during exercise.

For many, infection and disease eventually overcome. The body grows ill. The body dies. As in life, death has its own season. For victims of heart attacks, other cardiovascular ailments, and respiratory disease, the Grim Reaper comes calling most frequently in winter. Adverse weather conditions and cold temperatures have little effect on his rounds. Even in Hawaii, Death comes most often in wintertime. Residents in the southern hemisphere customarily get a collect call six months later.

Why this should be is unknown. Some have suggested that the seasonality of death follows some mysterious circannual disease susceptibility-resistance rhythm. Others have suggested that The End comes about due to a seasonal rise in the levels of particular hormones. Oddly enough, Reinberg and Lagoguey found that the levels of a number of hormones secreted by the adrenal glands, such as adrenalin and certain adrenal steroid hormones, rise during the winter months. These hormones influence everything in the body, from blood pressure to the immune system.

By then, though, the circannual rhythms and monthly cycles never really matter. We are faced with only one reality. In the end, death is the final season.

George Washington died in winter. The first President of the United States expired in his bed at his Mount Vernon estate on December 14, 1799. One day earlier, he had entered into his diary: "Morning snowing and about three inches deep . . . Mercury 28 at night." They were probably the last words he would ever write. The next day, he was ghostly pale and suffering, as one of his doctors, Elisha Cullen Dick wrote later, from "a violent inflammation of the membranes of the throat. . . . " Exhausted, nearly unable to speak, and fearing he would be buried alive, he whispered to his assistant to have him buried three days after he was dead. He died soon afterward. There was no struggle, no noise. The President had left quietly, without even a sigh.

On that same night in December of 1799, but miles away in a small New York town, an anxious Nathaniel Fillmore was giving comfort to his wife, Phoebe, now nine months pregnant. Cold and tired, she lay waiting to give birth to a new human life. Within her swollen belly, the future thirteenth President of the United States, Millard, was waiting to leave his mother's womb.

EIGHT

THE BODY IN YEARS

One morning, in the Peruvian autumn of 1939, a Quechua Indian woman and her daughter appeared at the door of the local doctor. The woman was in obvious distress for, as she explained to the physician, she was convinced that her child was possessed by demons.

The doctor examined the frightened girl. To his amazement, he noted that her abdomen was distended and swollen, a sure sign of advanced pregnancy. That this girl was carrying a child was surprising enough. But what really stunned him, and later astounded the scientific community, was that this pregnant girl was only five years old.

A few weeks later, on the fourteenth of May, a healthy baby boy weighing five pounds, thirteen ounces was delivered by Caesarean section from her womb, and the little Quechua girl named Lina Medina became the youngest mother in the world. To this day, she still holds that title.

The fact of Lina's pregnancy was left up to the local police to solve. But to the physicians and biologists who later examined her, Lina's sudden and early womanhood—she had her first menstrual period at eight months—remained the greater mystery.

Poor Lina's medical predicament, called precocious puberty, was a personal tragedy (in this case brought on by a tumor in one of her ovaries). Her childhood was abruptly ended by sudden adolescence, and on top of that, she had a young son to contend with (who later attended the same elementary school at the same time as his mommy).

Tragedy or no, Lina's plight brought into focus the idea that childhood can be shortened, say, by disease. Still, for most of us, the normal onset and duration of childhood, even adulthood, and old age are fairly immutable and fixed.

The cells lining the inside of the respiratory passageways of the windpipe and the lungs are covered by short hair-like devices called cilia. Cilia beat back and forth like oars moving the mucous and the particles of dust and dirt caught in it up to the throat at the rate of a half inch per minute.

Childhood usually ends when we are about twelve. It is followed immediately by a brief but turbulent adolescence, succeeded by a longer, more mature adulthood. Then, at some time in the future (usually around the age of seventy, give or take twenty years), when we are old and feeble, we drop dead.

Like humans, all other mammals (and indeed most animals that reproduce sexually) are born, go through a juvenile stage, reach sexual adulthood, get old, and then die. But, what differentiates us from other mammals, at least, is that we go through each individual stage for a much longer period of time.

It has worked well in our favor. The size of our brain and the remarkable expansion of the cerebral cortex is, in large part, due to our extended period of gestation and to our lengthy childhood. For most mammals, brain growth occurs only in utero. But, our rapid rate of prenatal brain growth has spilled into postnatal life.

Extending our brain development into childhood assures us of a long and wonderfully rich education. But, it has also left us utterly dependent on our parents for nourishment and protection. Indeed, a newborn cannot fend for itself or survive more than a day or so without its parents' intervention.

Intellectual development in return for absolute helplessness is a tough trade off. No doubt if we were like sharks, whose newborns are fully equipped with teeth, a hearty appetite, and the hunting skills of their parents, we might have been better off. But adult sharks have brains the size of lemons and all of the intellect of a toaster oven. Instead, to get us through our early years, we have evolved clever ways in which to gain nourishment, both physiological and psychological, from the one group that provides it: our parents.

One of the most striking observations in human development is that the body of a human infant looks different from its parents. An adult's face is more angular. The forehead slants backwards. The nose, and the jaw, big and imposing, juts forward. And the eyes, compared to the rest of the face, are relatively small.

Children, on the other hand, have large, round heads, and have big eyes compared to the rest of their face. Their jaws, too, are small. Moreover, their torsos are connected to short, stubby arms and legs. But the "Kewpie" doll appearance and adorable little mannerisms of children are there for an important reason.

As far back as the 1930s, animal behaviorists, like Austrian-born Konrad Lorenz, realized that animals respond to certain stimuli in their environment more than they do to other cues. A male robin, for example, will attack a bundle of red feathers more readily than it would a young robin lacking a red breast. The red feathers are the display of an adult male rival. A similar mechanism works for the herring gull. Herring gulls, which lay speckled eggs, will always sit on a more speckled egg rather than a less speckled one. For this bird, speckles are a sign of identification. In fact, the gull's trust in speckled eggs is so powerful, that it will stupidly sit on artificial eggs shaped like cubes and cones, as long as they are well-endowed with freckles.

A lactating mother produces about a quart and a half of milk every day. Even so, if milk is not continuously removed from a mother's breasts, the ability of the breasts to continue secreting milk is lost within one to two weeks. But, if the mother continues to be suckled, milk production can continue for several years.

The red breast of the robin and the spots of a herring gull's egg are called "releasing stimuli." And for animals, these special stimuli are particularly effective in triggering "innate releasing mechanisms," or special programs of behavior, in those they are intended for. Only a male robin will attack a red tuft of feathers because the threat display is meant for him. By the same token, only a female herring gull will witlessly sit on a speckled block of wood.

For humans, as Lorenz argued in his 1950 article, "Entirety and part in animal and human society," the precious and adorable little faces of children are the special stimuli that trigger (unbeknownst to us) "innate releasing mechanisms" in adults. Our reactions to these stimuli may not be as automatic as they are with other animals, but when we look at a child's big eyes and innocent appearance, we usually get an uncontrollable urge to pick it up and caress it. The more pronounced these "releasing" features are (the bigger the eyes compared to the rest of the face, for example), the more adorable the child, and the more likely we are to nurture it. In effect, cuddly infants produce more cuddling by adults.

This urge to nurture in response to a child's features, is so

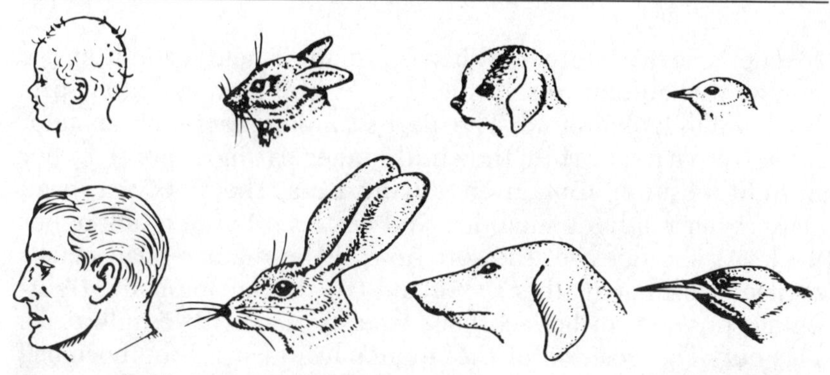

Human children are "cute" because their protruding forehead, large eyes, diminutive jaws, and soft, plump, rounded bodies trigger affectionate responses in adults. Adults find the young of most species "cute" for much the same reason.

generalized, Lorenz points out, that our behavior extends to the pet shop. Our love of baby rabbits, puppy dogs, and pandas, which possess big eyes, a bulbous cranium, and the clumsy movements of a young child, attests to this fact. We are drawn to them by the same mechanisms that lure us to our own children.

Hollywood has been well aware of our weaknesses for cuddly things. Indeed, when they want to make a creature from another planet appear harmless and sweet, they give it big eyes and a diminutive jaw. E.T. was made to appear adorable and cute by just this mechanism.

Certainly, "cute" is in the eye of the beholder. What is cute for humans, for instance, is not necessarily cute for say, monkeys. Indeed, monkeys have their own methods of relaying cuteness. Chimpanzee infants, for example, are born with pink faces. Their parents' faces are black. Infant rhesus monkeys are born with darker facial hair than their parents, and with a part down the middle of their scalp. They lose their dark coloration and the part in their hair in about six months, roughly at the same time that they are able to survive on their own without the care and nurturing of their mothers. Not coincidently, human children start to lose their "cute" appearance after a few years when they, too, can survive on their own.

Lovable looks, though, are only one tool a child has at its disposal to get it through the first few years of its life.

Mannerisms are another. When an infant is held upright at her mother's shoulder, for example, the child immediately lifts her head to look around. The alert scanning reaction is an automatic activity set off by the child's inner ear in response to her upright position. But after a baby tires, she puts her head against her mother's shoulder, and nestles it in the crook of her neck. As she does so, the soft fuzz of the infant's head makes contact with her mother's skin and triggers an automatic tightening sensation in her mother's breasts, and a little milk dribbles out. The position of the infant's head causes the hormone oxytocin to be secreted by the mother's brain. Oxytocin releases milk from the breasts. The stimulation is erotic and pleasurable for the mother, and it is rewarding for the infant. It reinforces nurturing.

Within 30 seconds after an infant begins to suckle the breast, milk begins to flow. The time involves nervous signals moving from the breast through the spinal cord to the brain. The brain then secretes the hormone oxytocin. Oxytocin travels through the bloodstream back to the breasts where it causes milk to be released.

A baby's expressions, too, are aimed to please, even though the child neither intends them nor is aware of the effects they are having on the mother. If, for example, the mother remains deadpan and still in front of her infant, the child will automatically go into a stereotypical behavior about three minutes later, designed to elicit a pleasurable reaction from her mother. At the onset, the baby will turn its head towards her mother and smile. But if her mother does nothing and continues to be deadpan, the child's expression will turn serious and her body will become still. She will then smile up at her mother again, but this time only briefly. Then, she will look the other way. In fact, she may run through this same pattern of behavior—smile, then sober up, then look away, then turn and smile—several times until the mother finally responds by caressing her. But, if her mother still persists in her deadpan expression, the child's face will switch from smiles to an expression of utter helplessness until, once again, the parent responds by nurturing her. As deliberate as the whole procedure may look to the adult, for the yearling, the exhibition is entirely unintentional.

Smiles may work wonders, but when they don't, infants pick out another arrow in their biological quiver to use in their quest for nourishment: They cry. Infants cry the minute they leave the womb (a reaction to their first breath of cold air), and use this behavior thereafter as a means of shutting out disturbing noises and as a means of discharging pent up energy. But this loud vocal behavior also is a powerful "releasing stimulus." It demands attention.

Sound is a potent "releasing stimulus" for the young of many species. Hens respond to the loud distress calls of their chicks, but not to their frantic movements. When frantically chirping young chickens are placed within a soundproof glass bowl in clear view of their mothers, the hens ignore them. But when the chicks are placed out of view and their hysterical sounds are allowed to be heard, their mothers react vigorously. By the same token, infant rhesus monkeys display a high-pitched "distress vocalization." When any rhesus adult hears it, they come running. The same happens when human infants cry.

If crying is a powerful "releasing stimulus" during the first few years of life, then so too is laughter. Laughter is generally regarded as an innate human response, for even deaf children laugh. It emerges in full bloom during the first few weeks after birth, although it is preceded by the smile by about a month. Usually, it is first observed after the development of smiling in infants as early as 5 to 9 weeks of age and is fully established by four months.

Charles Darwin, in his book, *Expression of the Emotions in Man and Animals,* written in 1872, suggested that laughter developed in early childhood because the long dependency in infancy required powerful signals that could reward caretaking by adults. Thus laughter, as a signal of well-being, developed as a counterpart to crying, a signal of discomfort. Indeed, when they laugh, children tend to produce sounds that are short and broken, which are, as Darwin observed, "as different as possible from screams or cries of distress."

Darwin, a student of human nature, was well acquainted with the interaction of children with their parents. And he probably noticed that children sometimes laugh or cry at the same situations. For instance, a baby might laugh when he is tickled, or when he is thrown into the air, or even laugh at the sight of a dog, or at the sound of a sneeze. But he may cry at these events, as well. To be sure, the events are stressful, yet laughter may be, in the long run, more adaptive than crying.

Some 15 years ago, psychologist Mary Rothbart at the University of Oregon, suggested that children manipulate their environment, and learn from it, by laughing at it. Her argument makes sense. If, she contends, a child were to cry each time it came upon a strange or unusual situation, an infant would most likely be removed from the disturbing surroundings and be comforted by its parents. But, in doing so, the infant would have little opportunity to become familiar with the situation or learn to cope with it. If, however, the child laughs, the parent assumes that the child is pleased and so tends to want to make the child laugh again by reproducing the disturbing situation. By repeating the stimulus over again, it gives the infant the opportunity to experience, to explore, and to learn more about the world.

No one knows whether or not our good-humored ancestors were able to adapt to their surroundings and survive better than their more serious siblings. But, no doubt, it certainly couldn't have hurt to laugh. Especially considering the problems children were up against. Just two centuries ago, for example, a horrifying number of children died in infancy and early childhood. Many of the surviving letters from the seventeenth century show the terror and panic of parents when their children fell ill with measles, colds, whooping cough, or the feared smallpox. In London, records show that between the years 1730 and 1749, seventy-five percent of the children born died before they reached the age of five. For this reason, women tended to bear as many as 12 to 15 children hoping that half would survive. "Twice five times suffered she the childbed pains," were the words inscribed on the tomb of a Scottish noblewoman who had died in 1676, "yet of her children only five remains."

Possibly because death came so readily to children, mothers used two childrearing practices to separate them emotionally and physically from their children, though they still took a

Charles Darwin, the nineteenth century naturalist, suggested that laughter rewarded caretaking in adults.

close and intimate interest in them. Almost as soon as they were out of the womb, the infants were wrapped up into little bundles of joy with linen or woollen swaddling clothes. And wet nurses were commonly employed to suckle infants.

The swaddling of infants, a custom dating back before the time of Christ, was designed not only to limit skin contact with mothers, but was also heralded in the sixteenth and seventeenth centuries as a method of preventing motion and of keeping the bones straight. Long strips of cloth were wound tightly around the child's legs. Another was bound firmly round his chest and arms. And still another band of cloth, known as a stay, went over the infant's head to keep it from wobbling to and fro. In this restrictive cocoon, the child remained for the first few weeks of its new life. Only when he was some four months old, were the arms freed. But he spent the first year of his life with his chest, belly, and legs bound in this cloth package.

Twins are born on average 19 days earlier than a single child. Their larger size stretches their mother's uterine muscle causing the muscle to contract that much sooner and push out the infants.

Being secured within this tight parcel was no doubt restricting for the child. It prevented the infant from grasping objects or exploring the world unfolding around him. But it was, at times, equally restricting for the mother, as she had to change these swaddling clothes when they became soiled, a procedure that often took an hour or more. Both mother and child were victims of the infant's slowly maturing nervous system.

Born with a nervous system that is not yet complete, a child's behavior is primitive; the infant has no control over its own bladder nor over its bowels, nor is it able to walk. Urination, defecation, and certain motions of the body are reflexive. Startle a child with a sudden clap of hands, and the infant will automatically throw its arms wide apart and then close them across its chest in an innate behavior known as the Moro reflex. Stroke its mouth, and the newborn will respond by sucking rhythmically.

The infant's behavior is "subcortical," a term based on the progress of cells which lay down a fatty substance known as myelin around nerve cells. Myelin, which constitutes the sheath of mature nerves, allows nerves to conduct signals faster. During childhood, its progress occurs in stages. Myelin reaches the lower regions of the brain first, then spreads slowly up to the cerebral cortex—from which originate the impulses for voluntary movement. At birth, however, myelin has only covered the lower, subcortical, brain regions and is why newborns rely on automatic reflexes. But, as myelinization progresses, these reflexive behaviors slowly disappear.

The slow advance of myelin matches the progress of voluntary muscle movement. By two months of age, a child can consciously move its arms and legs. By six months, it can raise itself on its hands and throw its head back. By nine months, it can stand up by holding onto a piece of furniture, and by a year to fifteen months, a child can walk on its own. Two years later, it can control its own bladder and bowel movements.

The steady progression of the laying down of sheath also coincides with the development of speech. The first sounds children learn are those that are the easiest for the mouth to produce. The sounds *pa*, *ba*, or *ma*, are produced in the front of the mouth and require the least tongue and air control. It is not coincidental, then, that the first word from a baby's mouth, spoken between the ninth and fifteenth month after birth, should be *mama*.

As emotionally charged as that word is when it is first voiced by a child, it is doubtful that many well-to-do mothers before the nineteenth century heard it. It was the fashion from the period 1500 to 1800 for wealthy mothers (and even many middle-class ones) to hire a wet nurse, a woman with breast milk whose job it was to suckle children. Often the practice was for entirely practical reasons; the mother herself had insufficient milk or was too weak to nurse. Sometimes, too, the husbands forbade their wives to breast feed, fearing that if they became pregnant again, they would not have enough milk for the existing child. But, the practice also had a biological function.

Lactation is a natural contraceptive. The production of milk induces a suppression of ovulation in women. Thus women who nurse are far less likely to become pregnant than women who do not suckle their young.

As a result, women who gave their children to wet-nurses were free of the birth-controlling effects of lactation. They could become pregnant almost immediately after they had their last child and could give birth to a new baby every year. Thus women who could afford a wet-nurse could bear 20, even 30 children (though many of them died) by the time their fertile years had ended.

There was yet another reason why women chose to have their babies fed elsewhere. Many mothers feared that suckling would scar their breasts. Indeed, it was fairly common during those years for breast-feeding women to lose their nipples completely because their older teeth-bearing infants chewed them off. In fact, it was the general custom in sixteenth and seventeenth century England to give an infant breast milk only until his front four teeth came in. After that, the child was weaned on bread, chicken, and eggs.

Adrenalin, a stress hormone secreted by the adrenal gland into the bloodstream, remains active for 10 to 30 seconds, followed by decreased activity thereafter for one to several minutes.

Offering up a tender breast to an infant with teeth is dangerous indeed. But, luckily for nursing mothers, most babies do not have teeth when they are born (although Julius Caesar, Louis XIV, and Napoleon were born with a tooth in their mouth). Normally, most babies get their first front tooth, the lower central incisor, about seven-and-one-half months after birth.

This first tooth is part of a set of 20 deciduous teeth, or milk teeth that erupt between the seventh month and the second year of life. But they don't stay around for long. Between the sixth and thirteenth year, they are slowly shed and are replaced by 28 to 32 larger permanent teeth.

A child's milk teeth do not erupt randomly, but usually develop successively, one type following after the other. First to arrive are an infant's four central incisors or front teeth. Then come the lateral incisors, on average, in about 13 months. Next to arrive are the first pre-molars about three months later. Roughly 19 months after birth, the conical and pointed canines erupt. Finally, the second milk pre-molars arrive in the rear when the infant is about two-and-a-half years old.

As it happens, boys usually develop their first milk tooth—the front central incisor—almost a month, on average, ahead of girls. Not all of a boy's milk teeth emerge a month before girls, however, but as a whole the development of the milk teeth is more advanced. For a given age, boys have more teeth in their mouth than girls.

The first permanent teeth, large, white molars (Latin for millstone) arrive by about the sixth year after birth but they don't replace any of the child's milk teeth. They simply rise up quietly in the rear of the mouth and stay there for the rest of our lives.

The first milk teeth to be replaced are the deciduous incisors, about a year or two after the large molars have erupted. As the permanent incisor tooth develops beneath the milk tooth, it pushes against the roots of the tooth above it. In response to this pressure and to the electrical forces generated by it, special bone-eating cells called osteoclasts dissolve the legs of the milk tooth in pace with the growth of the new permanent tooth. And the milk tooth falls out.

Meanwhile, without waiting for the completion of the first wave of permanent teeth, the front teeth are replaced in order back to the milk pre-molars in the rear. By the time the second pre-molar has snuggled up to the first premolar next to it, the child is about eleven years old. Roughly a year or two later, the second molar arrives. Finally, well into his late teens, the third molar, or wisdom tooth emerges in the mouth. Dental maturation is prolonged because there is a pause between in the formation of each individual molar crown.

In 1933, British physician H.L. Keane was not really interested in the teeth development of the seven year old girl in his office, though he made an entry in his notebook that most of her milk teeth were still present, with the exception of her first molars. He was more interested in the fact that this girl had the full, supple breasts, and the wide pelvis of a woman. Like Lina Medina, this girl had rushed into puberty early. And like the doctors who had examined Lina's condition, he wanted to know why.

Doctor Keane was not alone. The question as to what triggers puberty in the first place is on everybody's mind. There have been no shortages of theories. Some investigators suspect

Children often had wet nurses.

that the suprachiasmatic nucleus, the biological clock lodged in the base of the brain, is responsible for puberty. Others believe that the pineal gland, the cluster of specialized cells under our skulls sensitive to the rhythms of daylength, may help trigger puberty, as well. Still other researchers think that the final maturation of the brain along with a certain critical amount of body fat is important for puberty to occur.

Puberty in the human species is universal. We all go through it. A kind of metamorphosis, puberty molds the young, juvenile body into a sexually mature adult. Girls suddenly discover boys. Boys suddenly discover *Penthouse.*

In the male, the testes and penis grow larger. Coarse facial hair develops and pubic and underarm hair appear. The voice deepens. The muscles enlarge. The limbs elongate.

There is a variation on the same theme in girls. Here, though, puberty prepares the body for child bearing. In the process of puberty, girls begin to notice the growth of their breasts and the maturation of their genitals. Their hips grow wider. And, like their male counterparts, they too grow coarse pubic and underarm hair. Puberty in girls also leads to the beginning of their first of many menstrual cycles, an event called menarche.

In the United States, the development of these secondary sex characteristics begins around eleven years of age in girls and six months later in boys. Girls complete the process of puberty on the average in about four years, boys in about three-and-one-half years.

While the timing of puberty seems to be primarily determined by our genes, scientists are less interested in exploring our DNA than they are in studying the processes in the brain, for it is here that puberty begins. Melvin Grumbach, at the University of California, believes that puberty, in fact, is not so much triggered as it is held back and released—but only when conditions are "right" for parenthood. Grumbach uses sheep as a model in his study into the neural and hormonal mechanisms behind puberty. Sheep make particularly good specimens in the study of human puberty because they seem to reinitiate the event every year.

A swallow takes one to two seconds. It involves the coordinated activities of the tongue, the vocal cords, the epiglottis, and the esophagus.

According to Grumbach, puberty is not so much an isolated event as it is a link in a whole chain of events that have their origins in the womb. In the embryo, as well as in the adult, the small mass of brain tissue called the hypothalamus, directs the activities of many of the developing tissues of the body by secreting special "releasing factors." One factor in particular, gonadotropin releasing hormone, or GnRH, is produced in the human hypothalamus as early as the eighth week of gestation. By mid-gestation, this protein substance trickles down to the body's master gland, the pituitary, causing it to release gonadotropins, other hormones that in turn trigger the release of sex hormones. The ovaries produce estrogen. The testes secrete testosterone. Like the powerful messengers they are, the sex hormones from the gonads then travel back to the hypothalamus causing it to release less GnRH, a mechanism called negative feedback.

In the first few months of life, babies produce sex hormone levels as high as those of adolescents. But the secondary sex characteristics of the body don't develop because GnRH secretion soon decreases and for some years thereafter no longer stimulates the pituitary gland. The whole system, in effect, shuts down. No one quite understands why. But it remains in this quiescent state for nearly ten years.

But, then, something happens. Puberty begins. What flips on the puberty switch is still unknown. Grumbach suggests that the hypothalamus is held back from triggering puberty because of two inhibiting mechanisms. The first mechanism Grumbach proposes involves the idea that the hypothalamus's sensitivity to the gonads' sex hormones changes over time, being super-sensitive in the beginning and later less sensitive to their influence. Grumbach's second mechanism postulates that the brain itself is capable of shutting things down and then, when conditions are right, of turning things on.

The overly-sensitive hypothalamus idea is supported by the observation that when minuscule amounts of sex steroids are injected into children, the secretions of gonadotropins are abruptly shut off. Grumbach suggests that this means that if small amounts of steroid hormones from the gonads can shut down the hypothalamus, the hypothalamus must be highly sensitive to any kind of gonadal influence.

The life time of a red blood cell in males is about 120 days; female red blood cells die on average eleven days earlier.

But researchers have also noticed something else. In children with gonadal dysgenesis, individuals who lack functional gonads, there is a dramatic fall in the level of secretions of gonadotropins by the pituitary between the ages of four and eleven, the period when the hypothalamus normally is quiescent. Somehow, without the intervention of the gonads' steroid hormones, the brain can shut itself down.

Putting the picture together, Grumbach and other researchers think that during childhood, the hypothalamus is inhibited from secreting large amounts of its GnRH either because it is held back from doing so by the higher regions of the brain, or because it is insensitive to the calls of the sex hormones. In any event, for most of the period of childhood, the hypothalamus remains dormant. Then suddenly, the hypothalamus is reawakened. It becomes far less sensitive to the inhibition from the gonads and begins to pump out large rhythmic bursts of GnRH. The increased flow of GnRH stimulates the pituitary to dump out greater amounts of gonadotropins which, in their turn, wake up the gonads. Estrogens and androgens flood the circulation and puberty rolls on.

What exactly gets the puberty ball rolling isn't as yet known. But Grumbach suggests that it can occur at any time. Lina Medina was just one famous example. But doctors have seen infants with precocious puberty—mature genitals and so on—dropping right out of their mother's womb. The mechanism isn't perfect.

Ironically, the fundamental apparatus in humans seems to be simpler, perhaps more primitive, than it is for other species. Seasons and other environmental influences appear to have little effect on us.

We are not completely immune to environmental cues, though, just less sensitive. Spring fever is one phenomenon that everyone mentions. For the college crowd, this hormone-popping event seems to occur as soon as the cold days of winter are over, an indication that the lengthening daylight hours may have an effect on our reproductive behavior. Then there are the curious cases of synchronous menstrual cycles whenever groups of women live together for an extended period of time. For some strange reason, roommates often wind up having their periods on synchronized schedules. It is possible, then, that some as yet unknown pheromone, or airborne body scent, does indeed influence our reproductive lives.

Puberty is not an immutable process; it can be arrested or even reversed. Smoking marijuana can delay the onset of puberty. So can excess dieting. Rose Frisch, at Harvard University, argues that a critical amount of body fat is needed to satisfy the demands of puberty and also to sustain menstruation. Such changes are well-known to endocrinologists looking into the effects of vigorous physical exercise on already thin young women such as ballet dancers and cross-country runners.

Girls who start their strenuous physical activities, and lose body weight accordingly, before their first menstrual period are the most likely candidates to have their entire puberty train halted. Those who do start this exercise program early, risk having to start puberty much later than the other girls their age. But for some investigators, just simple weight loss does not seem to be the culprit. Studies have shown that in girls who cut down on their sports activities but do not gain weight, puberty occurs on time anyway. Whether it has to do with the energy requirements of the body, no one knows. But, if some physiological threshold is not reached, puberty does not begin. It's almost as if the body is saying that it's not a good time to get pregnant.

While many factors seem to play a role, nutrition may also influence the timing of puberty. In areas where nutrition is poor, for example, there is a delay in the onset of puberty in children. The theory, according to some biologists, is that good nutrition and good general health are necessary if the genes responsible for puberty are to become activated. Those children who have poor diets or a history of long illness reach puberty at a later age. As it happens, in many parts of the Third World, rural girls reach puberty much later on average than do urban girls who are presumably healthier than their country cousins.

Such is life in the twentieth century, but the onset of puberty today occurs earlier than it has ever occurred in history. There has been a long and well-documented trend towards earlier puberty. In Bach's time, around 1740, the average age at which choir boys cracked their voices was about eighteen years of age compared to about thirteen years today. Medical writers during Europe's medieval period commonly placed a woman's first menstrual period at between twelve and fifteen years of age. The medieval Church of Rome set the age of puberty and of marriage at twelve years for girls and fourteen years for boys.

The medieval canonical requirement was that girls be at least twelve years old at the consummation of their marriage. In 384 A.D., thirty-year-old Augustine of Hippo (later Saint Augustine) plotted to wed a girl "two years below the marriageable age." She was probably no more than ten. "I liked her," he wrote in his *Confessions*, "and was prepared to wait."

From the fourteenth and fifteenth centuries on, rich urban girls tended to be very young for their first marriage, younger in fact than girls of lower social order or even country girls. The wife of Chaucer, a middle-class girl of the city, was first married at age twelve. Many city girls were mothers by the time they reached their fifteenth birthday.

For medieval women, nutrition seemed to play a great role. With better diets, the rich urban girls experienced their first menstrual periods around thirteen years of age. Peasant girls living in the countryside, and those of lower social station seemed to achieve menarche closer to age fifteen.

Whether girls and boys entered puberty together early in our history is not entirely known. But one current observation appears to be nearly universal: Girls today enter puberty earlier—about six months earlier—than do boys. That girls reach puberty sooner than boys has dramatic consequences on both physical and psychological development. Growing up, I remember how "mature" the girls my age were and how, by comparison, my male friends behaved. The girls were having crushes on their teachers. We were still fighting over baseball cards.

Strong mixing waves of the stomach come once every 20 seconds.

Puberty timing plays a role in height, as well. Boys are, on average, taller than girls. In fact, by the time puberty begins, girls have already reached the peak of their adolescent growth spurt. By the time of their first menstrual period, girls are already slowing down. Boys, on the other hand, are just beginning to reach their growth stride by the time puberty approaches. So boys grow taller because they have more time to do so, while girls are left behind looking up.

Why this happens probably has to do with the fact that male brains and female brains are wired differently and are influenced by different types of hormones. In girls, for example, estrogens made by the ovaries, are secreted earlier than the male sex hormone secreted by boys. The result is that the female sex hormones reach a more potent physiological level in girls more quickly than the male sex hormones do in boys, causing puberty to arrive earlier.

What causes these estrogen hormones to rise so early in girls is not known. But this early rise may have a selective advantage. Girls may need that extra time to grow and organize the apparatus necessary for child bearing.

As the mystery of puberty becomes unraveled, it is apparent that we still know very little about the mechanism behind this once-in-a-lifetime event. As investigators in this field will drolly point out, research into puberty is still in its own early adolescence.

Hunger pangs in the pit of the stomach begin some 12 to 24 hours after the last meal; in starvation, they are strongest in three to four days and then gradually weaken thereafter.

When James I of Scotland decided in 1428 to marry off his eldest daughter, Margaret, to the heir to the French throne, she was only eleven and had not yet reached the age of puberty. In those days, childhood was a brief moment in the lives of girls. And it was followed immediately by adulthood. There was no adolescence. Boys quickly became men. Girls became women. Men went to work and to war. Women bore their children.

This was, of course, what King James I had in store for his child as he had her shipped off to Tours, France. There, she married the thirteen-year-old Dauphin. But the marriage ended tragically. Too immature and too hostile to appreciate the Scottish Princess, the French heir neglected her. Thus ignored, young Margaret spent the rest of her life reading literature and writing poetry before dying unhappily at the tender age of twenty.

Like many women in Medieval Europe, Margaret did not have a chance to bear children for her husband nor did she, like many of her time, see old age. But, had she reached the age of fifty—which was unlikely in those days—she would have come upon an unusual event in the biology of the human species, an event called menopause.

Menopause—the complete cessation of the monthly menstrual cycles—is a fairly recent event in human history, mostly because the chances of a woman surviving to the age of menopause were relatively low until the beginning of this century. Between 10,000 B.C. and 1640 A.D., most women could expect to live to about thirty-two years of age. Complications in childbirth and infectious diseases took their toll. By the beginning of the nineteenth century, an average woman's life expectancy was still only about 50 years: The age typically marking her passage through menopause.

Menopause is not an abrupt termination of a woman's reproductive life. Rather, it is a transitional event, one that may span 10 or even 15 years. Usually it starts anywhere from the age of 45 to the age of 53. There are, however, a small number of women who begin menopause in their early 40s and some who begin after the age of 53, a fact that seems to have been known for some time. "[Women] have bleeding which makes their bodies clean and whole from sickness. And they have such purgations from the age of twelve to fifty," were the words written in the fifteenth century *Sloane Manuscript 2463*, the first textbook of gynecology published in English. "Even so, some women have purgations for a longer time. . . . "

No doubt, women have it tough. And no doubt, men have it easier. By the time a man has reached the age of 45, the amount of testosterone secreted by his testes is significantly lower than it was twenty years ago. By the time he is 60, the level of this potent male sex hormone in the blood is close to what it was before puberty. But even in old age some testosterone is produced. And, although the quantities of the hormone are small, a man still produces enough testosterone to keep his beard growing. He also produces just enough of this male sex hormone so that it is possible for him to father children later in life, when he is sixty or even seventy years of age.

However, a woman's ability to bear healthy children gradually lessens as she reaches middle age until, finally, she cannot bear any children at all. In a real sense, menopause is the flip side of puberty.

During and after menopause, the human ovary stops producing estrogen hormone. Without estrogen, the body parts maintained by this hormone eventually decline. The vagina loses much of its thick inner lining and shortens. The breasts become flabby; the bones brittle. "Aging spots" and pale "bleached" areas appear on the skin. The skin wrinkles. Sweat glands atrophy making the body more sensitive to temperature and humidity changes. The ovaries shrivel.

Menopause also brings on other psychological and physiological problems as well. Insomnia, night sweats, anxiety, depression—all come about, in varying degrees with time.

The vibrations in the air constitute sound waves. The higher the pitch of sound, the greater its frequency or cycles per second. Adults can detect sounds that have a frequency between sixteen to about twenty thousand cycles per second, but hear best at frequencies ranging from one thousand to two thousand cycles per second. Children hear higher-pitched notes than adults. After puberty, however, the sensitivity to higher notes declines at the same time that the voice deepens. Our ears, then, are best adapted to the pitch or sound frequencies of human conversation.

"Hot flashes" occur, too. While their cause is unknown, hot flashes are characterized by sensations of heat, usually in the face, neck, and chest and are sometimes followed by perspiration or shivering. The average duration of a hot flash is about three minutes and usually starts in the head, neck, and ears, although many women experience them all over.

Perhaps, because of these unusual biological changes in the human body, older women in the past were not treated in the best of light. During the Middle Ages, for example, a woman who had passed the age of menopause was, in the literature of the time, usually portrayed as a witch or a cunning temptress. "More than all of them I despise," John Gower, a contemporary of Chaucer in the 1300s, wrote in his book, *The Mirror of Man*, "The old woman who is flirtatious / When her breasts are withered." Which is probably why Gower married a woman a third his age—his second wife—when he reached the ripe old age of sixty-eight.

The sixteenth and seventeenth centuries were no better for the older single or widowed woman. She was still treated as an object of satire and scorn. In William Shakespeare's play, *The Taming of the Shrew*, a woman is described this way: ". . . an old trot, with ne'er a tooth in her head although she may have as many diseases as two and fifty horses."

If the history of public opinion has changed over the years concerning menopause, so too has the history of biological opinion concerning this female event. But opinions as to the cause of menopause fall into two basic camps.

When a woman is born, all of the eggs she will ever use in her life are packed into her two ovaries. As she matures, the eggs are tossed out once a month during each of her over 400 menstrual cycles. Many more of them simply disintegrate. Eventually, she runs out of eggs. So, one opinion goes, no eggs, no children, no menstrual cycles. Menopause.

Well, not entirely, according to the other opinion. True, the store of eggs within a woman's ovaries is nonrenewable. But, menopause is not entirely due to the "wearing out" of these reproductive organs. There is abundant evidence that other factors besides the ovary are at least partly responsible for the halting of egg release.

If, as the argument goes, an aging set of ovaries limits the period of a woman's reproductive life, then one would suspect that if these old ovaries were replaced by younger ones, their reproductive function would continue. In rats, at least, that doesn't seem to happen. The first experiments with rats, in fact, showed that senile animals did not return to their normal egg-releasing cycles after their ovaries were replaced with younger ones. Yet, when these old ovaries were placed in the bodies of young rats, the aged ovaries seemed to continue to be healthy and active for many more months. So in rats, anyway, the answer seems to be that both the ovaries and the hypothalamus seem to be at fault.

In support of their argument, the egg-loss-isn't-everything group makes the point that although the age of onset of menarche occurs at a younger age now than when it occurred, say, 500 years ago, menopausal age appears to be more firmly fixed. Studies of old records have shown that the average age of menopause has not changed much since antiquity. Indeed, there seems to be no relation between the age of menarche and the age of menopause.

The present record bears this out. Women who start their first menstruations late, for example at 16, are likely to end their reproductive period first, while those who start early are likely also to continue menstruating well into their 50s. In fact, there is often a strong family pattern in the timing of the menopause; those whose mothers and elder sisters finish their reproductive cycles early are likely to finish early, too.

Surely, if a girl possesses a limited supply of eggs when she is born, and then starts to release them early in life through her monthly menstrual cycles (as well as through disintegration), and *still* has her menopause at 50, the loss-of-eggs theory must not be entirely the cause of this event. After all, look at the case of Inacia da Silva.

Born in Brazil in 1877, Inacia had her first menstrual period when she was just eight days old, and (her mother said) menstruated regularly each month thereafter. Then, one day her parents noticed that Inacia's abdomen was larger than it should have been and rushed her to the local hospital. The doctors said she was pregnant. They were right. Four months later, she delivered twin boys, both dead. She was just seven years old at the time. Years later, at the age of 50, Inacia experienced her menopause.

Hunger contractions of the stomach, when it is empty, last some two to three minutes.

When doctors last checked up on her in 1947, Inacia da Silva was still alive. She had just celebrated her 70th birthday which, for most people these days, is not uncommon. Indeed, a man in the United States can expect to live nowadays approximately 71 years; a woman 78.

Stacked against the life spans of many of the other creatures sharing this planet with us, that's not much. A mighty sequoia lives to be more than 3000 years old. A young bristlecone pine planted today can live to see the year 5988 A.D. Still, a squid lives only four years. A mayfly, up at dawn, lives only until sundown.

There is no doubt that the average human life expectancy has increased tremendously since the time of the Roman Empire when a person could expect to see only the age of 25. Through better medicine, a reduction in infant mortality, vastly improved sanitation, and other methods of controlling infectious diseases, doctors have managed to bring life expectancy up to where it is today. But what they haven't been able to do is to change the length of the maximum life span of the human species. That has remained fairly constant throughout history. Indeed, once a man reaches the age of eighty, the prospect of living much longer than that is only a little better than it might have been in the seventeenth century.

Certainly, there are those among us who have lived longer than the expected date. The Greek poet, Sophocles, was in his early nineties when he died. Picasso was 91. In Sweden and in the United States, where careful records are carried out, a few persons have exceeded 100 years of age. According to the Bureau of the Census, there will be more than 4 million persons over the age of 85 by the year 2000. But they are, in the words of the American poet Archibald MacLeish, "the last leaves on the oak."

Much of the confusion about persons living as long as Galapagos tortoises, has come from the press. They get their stories from the claims of people from remote places such as Soviet Georgia. People there often report living to the age of 160. But there is no proof of them living nearly that long. Their apparent longevity is not due to good eating and clean living, but to poor record keeping and illiteracy. The problem was best expressed in a Sidney Harris cartoon where one elderly villager explains to a young visitor, "Our reputation for longevity is based on several factors: hard work, simple food, lack of stress, and the inability to count correctly." According to the *Guinness Book of World Records*, the longest anyone has been able to live has been to the age of one-hundred and fourteen.

Judging from the volume of material out there, many scientists believe that human age time limit is built into our bodies the way planned obsolescence is built into light bulbs. Stanford University internist, James Fries, has gone so far as to say that there is such a thing as "natural death"; people will die around the age of 85 even after man learns to cure every disease under the sun. To be sure, his statement provoked a lot of criticism from his peers. Still, when pathologist Robert Kohn, of Case Western Reserve University, recently performed autopsies on 200 people 85 years and older, he and his team could not identify a cause of death in thirty percent of them. Evidently, these people had died of old age; they had reached their natural genetic limit.

There are 10 billion capillaries in the human body. It is in these vessels that red blood cells give away the oxygen they carry to the other cells around them. Nevertheless, a circulating red blood cell only remains in a capillary for 1 to 3 seconds before it heads for the heart again.

That our bodies should have a programmed death is based in part on the work of microbiologist Leonard Hayflick, of Stanford University School of Medicine. In 1961, Hayflick discovered that cells possess their own biological clock, a timepiece that schedules a cell's ultimate lifespan. For several months, Hayflick raised normal human fibroblasts in glass flasks. Fibroblasts are actively dividing cells that build cartilage and other connective tissues of the body. Fibroblasts grown in this manner divide regularly until they coat the entire surface of the flask. The microbiologist then transferred the cells to two flasks and let them grow once again until they covered the glass container. Each transfer, called a doubling, produced the same results. And when these flasks were covered, Hayflick transferred them again.

For twenty-five, even thirty doublings everything went fine. But by the 35th doubling, the fibroblasts were dividing more slowly than before. Eventually, by the 50th doubling, and some nine months later, the cells stopped dividing altogether and died; they had reached the genetic limits of their life span.

Next, Hayflick repeated the experiments, with human fibroblast cells from elderly donors and with cells from human embryos. Surprisingly, the fibroblasts from the old folks did not divide as much as the cells from the embryos did. The number of doublings were somehow related to the age of the individual.

Hayflick's discovery was an eyeopener for those who still clung to the belief that there was no time limit to the human life. But, what made it all the more convincing was that the effect was nearly universal in animals. Fibroblasts removed from the embryos of mice, which live only three years, divided about 15 times before they died. Fibroblast cells taken from Galapagoes tortoises (whose life span is about 175 years) divided roughly 90 times before they met their grim reaper. Our fibroblast divisions, of course, fall somewhere in the middle of both of these animals, at 50 divisions. But then, (according to James Fries) so does our maximum life span.

The biological clock is located in a cell's nucleus, the part of the cell which contains its genetic instructions, the DNA. If a nucleus is removed from a young fibroblast and placed into a senescent fibroblast, the newly formed cell thinks that it is a young cell and not an old one and doubles accordingly.

Still, tearing cells from their natural hosts and then watching them slowly die in a dish seems unnatural. After all, normal cells don't live in glass houses. It begs the question as to whether or not cells, left to themselves, would also die of natural causes, too.

In 1981, two University of Georgia botanists, Jeffrey Pommerville and Gary Kochert, were probably asking themselves that same question. They were studying the tiny green pond creatures called *Volvox carteri* in an effort to understand why these little beasts divide a number of times and then ultimately die.

Up close under a microscope, *Volvox* appears as a small, clear green hollow ball or spheroid, surrounded by cells. Each one of the 4000 or so cells has two tiny whiplike tails, flagella, embedded in them which they flail around in somewhat coordinated unison to move the spheroid about. Each cell of the spheroid harbors minuscule organelles called chloroplasts that provide the creature with food. But *Volvox* also has 8 to 16 other cells, called gonidia, located in its bottom half. The development of the flagella-possessing cells begins when each gonidium within a parent spheroid divides. After a number of divisions, the gonidium forms a daughter *Volvox*.

Provided with a means to make food from sunlight, *Volvox* would appear perfectly designed to live forever. Yet, it doesn't. It lives for only a week. Its rapid agonizing death is, in a sense, a vision of our own aging.

At three days, the *Volvox* is still young and the picture of health. It swims about its freshwater universe, its thousands of flagella whipping vigorously to and fro. But by four days, it is already showing wear. Small globules of fatty lipid appears in each of the spheroid's cells. Finally, by the time it has reached the end of a week, it is shabby and tattered. Its chloroplasts, once ripe and healthy, now appear haggard and gaunt. The flagella now barely move. Within hours it is dead. Its brothers and sisters die the same way and over the same period of time.

Exactly how *Volvox* died, isn't really known, although the two botanists suspect it was from starvation due to old age; *Volvox* was just too tuckered out to make food. But it also may have been because *Volvox* was simply programmed to die. Indeed, one theory proposes that aging is just one frame of a larger sequence of events in the life of an individual.

Chyme (made from digested food and water) moves along the small intestine, on average, at the rate of a quarter of an inch per minute. At that rate, it takes some three to ten hours to travel from the stomach end of the small intestine to the entrance of the large intestine.

It has long been known that certain genes control the growth and development of the embryo. So why shouldn't the body possess genes that program the aging process? The theory goes that old age is simply the result of the cells of the body reading their DNA from cover to cover. In a sense, cells would read the first chapters during conception, and the last ones during senescence. The problem, however, is that for the process to work (that is, for it to be passed down from generation to generation) it would require that our ancestors live until their natural senescence. Most of them, of course, did not.

Left undisturbed, our cells might be able to read this genetic literature in peace. But the body is constantly in a state of flux, bombarded as it is from the assaults of the environment. Ultraviolet light and pollutants take their toll. So the body ages, not from the constant wear and tear, but says another theory, from its failed attempts to repair the damage done. At first the DNA is mended. But, gradually, the errors in duplicating its sequence each time a cell divides catches up with it. The errors accumulate, abnormal proteins are made, and the body's cells can no longer function properly. And so we age.

Whatever its cause, programmed senescence or the failed repair of environmental molestation, the normal aging process offers a bleak future. Wrinkled skin and lost hair, age spots and hearing loss are some of its symptoms. But the brain, too, slowly deteriorates, beginning at the age of about 25. It shrinks. By the age of 70, a person's brain is some eleven percent smaller than it was when he was 40. The shrinkage is accompanied not just by the loss of nerve cells in some brain regions, but also by the degradation of the parts of some neurons and a drop in the levels of some of the neurotransmitters, the chemicals that carry signals between cells.

Living into our seventies and eighties, we have traded off the acute childhood illnesses of the past—smallpox, polio, and pneumonia—for the chronic illnesses of the future: arthritis, cancer, heart disease, and Alzheimer's disease. We will live longer, but not always in good health. By the year 2000, the number of old people living in nursing homes is expected to rise by 54 percent.

If the whole aging process is so terrible, why do we go through it? One answer may be that the aged are important for the continued care, nurturing, and survival of our slowly developing children. Since the juvenile period for humans lasts for many years, it is advantageous (in an evolutionary sense) for there to be adults around to take care of the young and to teach them the ways of the world. So, our longevity, then, is for the child's sake. He needs us to stick around until he can take care of himself. Indeed, it's not uncommon nowadays for women to live nearly twice the age of their menopause. But eventually we reach our nadir.

And death arrives.

Down here, the occupants don't rest in peace. They are probed, tagged like furniture, and shoved into tiny, refrigerated closets too small to sit up in. But they don't complain. This is a hospital morgue. And these people are dead.

For most it will be a temporary stay. Soon their relatives will come by to claim them. But the other ones here will be held a while longer. Victims of foul play or runaway cancer, their corpora await the coroner's knife.

In the terms of the living, time here has ended. Hearts have long stopped beating. Chests no longer heave. Now only decay and decomposition set in. Salts, no longer held within cells by energy-expending pumps, flood into the blood. Hormones, trapped in glands during life, now ooze out into body cavities. Fluids settle. Blood clots. The body cools.

Fecal waste matter (which is 30% dead bacteria, 20% fat, 20% inorganic matter, and 30% undigested roughage of food) is pushed through the large intestine towards the anus by propulsive muscle movements. These movements occur only a few times each day. They are most active for about 15 minutes during the first hour after breakfast.

To the specialists who make their living studying the corpus postmortem, death follows its own lonely drumbeat. Knowing this, it is possible for a forensic scientist to count backwards from when he discovered the body and estimate the time of death. Estimate, because death time does not proceed in a set and predictable fashion. As every researcher in this business will sadly relate, there isn't a formula yet that can precisely determine the exact time of death.

Back in the days when the technology to uncover such things was far more crude, defining when a person had died was at best cautious business. For years, doctors listened attentively for the telltale lack of a pulse or, failing that, the absence of a breath, usually determined by holding a cold mirror in front of the nose to detect the expiration of moist air. These simple methods were not always foolproof. Sometimes a person, found to be dead, would rise from bed, frightening the children. The family unexpectedly had to cancel the funeral plans.

This fear of the dead returning to the dinner table probably led to the practice of burying the loved one underground or, as is the case in some countries, of breaking the leg bones of the deceased or binding its hands and feet. Supposedly, the dearly departed would be unable to rise from the grave and haunt family and friends if it were unable to walk. The Mafia have another trick. They bury their victims face down. A lot of the Mob's ghosts have ended up in China that way.

Defining exactly when a person is dead came much closer in 1950 with the landmark case of Thomas vs. Anderson, argued in the California District Court of Appeals. At that time, it was determined that death "occurs precisely when life ceases and does not occur until the heart stops beating and respiration ends." Later, the legal definition of death went further to include brain death, as well. Death of the brain occurs when the electrical activity of the brain no longer is visible indicating that the cells of the brain have expired.

This viewpoint is by no means defunct today, but a number of scientists would argue with it. For one thing, not all of the cells of the body die when the owner does. Many of the cells of the muscles, for example, continue to live, sometimes for a few hours, after a person has been declared legally dead.

This peculiar fact of death can be demonstrated with electrodes placed into the skeletal muscles of the legs and arms. A coroner, with a little jolt of electrical current, can trigger the muscles of the limbs to contract, often until one or two hours after the death of the individual.

The most bizarre demonstration of this phenomenon occurs when electrodes are connected to the muscles of the face. For the novice it is at once fascinating and unnerving to watch the face of a corpse already a half hour dead, wince, frown, and squint as the current is applied to the muscles surrounding the eyes and mouth. But this strange occurrence eventually disappears after an hour and a half as the muscle cells themselves run out of energy and die and the face remains literally deadpan.

Scientists take this and other factors into account in their quest to uncover when the deceased died. There are signs to look for and events to study. Death is not static. Neither is it simple.

Researchers in this line of work point to the fact that the time, onset, and rate of body decomposition varies according to such factors as wind, humidity, outside temperature and even the disease state of the body before it died. Their job is to factor in all of these variables and reasonably come up with an estimation of the time of death. It is seldom easy.

One forensic pathologist points to an incident which occurred a few years ago. A boy had stabbed his parents one summer day. Several days later, the bodies were discovered by the police. The boy's mother was found in the cool basement with little decomposition evident. The boy's father was discovered in the upstairs bedroom where the temperature had reached 90 degrees for several days. Unlike the body of the woman, the father's body was blackened and bloated with signs of massive decay. And yet, as subsequent evidence revealed, both parents had been killed within a few moments of each other.

When the environmental conditions are controlled, as they are in the laboratory, certain events can be seen to occur immediately after death. Many of them are strictly chemical in nature and proceed at standard rates. The levels of sodium and chloride, for example, fall in the blood and in the fluid bathing the brain and spinal cord, and potassium levels rise after death. This curious phenomenon occurs as cells, which normally hoard potassium and actively keep most of the sodium and chloride out, run out of energy and die. When death comes, potassium floods out of the cells. Other salts wash in.

Other chemical changes are more obvious to the observer. Soon after death, for instance, the muscles of the body begin to lose their tone. By 2 to 4 hours after death, advancing to about 12 hours, muscles become rigid and stone-like. The process is due to the lack of energy for the muscle cells. Without the power to generate contraction, the protein fibers that are responsible for muscle contraction lock together like pieces of

Velcro. Forensic pathologists have a name for this event. It's called *rigor mortis*.

Coroners and police detectives alike use the degree of muscle rigor and the posture of the body to estimate the time and place of death. As rigor sets in, the body conforms to the shape of the object beneath it, and freezes in place. No doubt that if a body is found lying on its back with its legs pointing towards the sky, it probably became rigid somewhere else and was transported to cover up a crime.

Not surprisingly, superstitions abound in regard to the effects of muscle rigor. One of the most common is that it causes the bodies of the dead to sit up in their coffins several hours after they have expired. This is not true. Rigor hardens the muscles, not contracts them. Bodies are unlikely to sit up once rigor has set in. Interestingly, *rigor mortis is* responsible for the gooseflesh seen on corpses. Called *cutis anserina,* it occurs due to the rigidity of the short muscle fibers in the skin which attach to the skin hairs.

Once full *rigor mortis* has been achieved, it lasts for a few hours then slowly disappears. With time, the muscles once again become soft and the limbs are again flexible. The duration of this rigidity is often variable, influenced by both the temperature of the body and the environment. During the hot summer months, rigor may vanish some 9 to 12 hours after death.

From the investigator's standpoint, muscle rigidity is just one clue to look for in determining the time of death. *Livor mortis* is the other. Appearing as a purple discoloration of the skin caused by the settling of the blood as circulation stops, *livor mortis* formation begins immediately after death. After two hours, large royal blue blotches are perceptible on the bottom of the corpse. This purple discoloration becomes even greater between 8 to 12 hours after death. Forensic scientists have long noted that sometime after 12 hours, this lividity becomes "fixed" and doesn't shift even when the body is moved. Investigators use this fact as evidence to indicate whether or not a body has been placed elsewhere after death.

Muscle rigidity and blood settling are difficult events to manipulate. But altering the rate at which the body cools has always been the trick of mystery novels. Corpses have been stuffed into freezers, submerged in cold water, even packed in ice in an effort to throw off the detectives. Actually, though, the novels have been using old data.

It used to be assumed that the body cools some one-and-a-half degrees an hour. And, indeed, forensic scientists of old used to use this figure as a rule of thumb in judging the time of death. But as one scientist put it, "this now seems to be correct 50 percent of the time." Today, with more sophisticated analysis techniques, the rate of cooling of the body as it slowly reaches ambient temperature, is far more complicated. During the first hour, researchers now know, little cooling of the body occurs. After that, the rate of cooling is influenced by everything from the temperature of the surroundings, to the wind and humidity of the area, to even the amount and type of clothing the corpse happens to be wearing. Needless to say, it sometimes takes a computer to figure in all of the variables. And even then, other facts are necessary.

Examining the dead is a messy business. What makes this job more unpleasant are the foul odors that emanate from the body as decay and decomposition set in. There are two parts to decomposition: autolysis and putrefaction.

Autolysis occurs when the digestive enzymes in the stomach and small intestine break down the walls of the gut and pour into the body cavities. Still active and uninhibited, they digest other organs until they themselves are broken down. As a result of this activity, coroners often find perforations of the stomach within a few hours of death.

The ability of the brain to detect the location of a sound depends on the differences in the time of the arrival of the sound to the two ears. We can detect the source of a sound even if it arrives in one ear a hundredth of a second before it gets to the other ear.

Putrefaction is by far autolysis's uglier sister. We may die but our bacteria live on. As death progresses, tissues are digested into slimy fluid and foul-smelling gases are given off. As a result of the gas released within the tissues, the body becomes distended, often bloating to three times its natural size. The skin changes color from green to black. The eyes and tongue protrude from the pressure. What's more, the upper layers of the skin softens and slips off.

And then there are the insects.

It doesn't get any prettier. By 12 to 18 hours after death, decomposition is so advanced that the features of the corpse become unrecognizable. After a few days, the body begins to dry. The skin hardens to a paperlike consistency. The tips of the fingers and the skin of the face shrink, giving rise to the false belief that hair and nails grow after death.

Much, much later, only the bones will be left. Then, ashes to ashes. Dust to dust.

NINE

TURNING BACK THE CLOCK

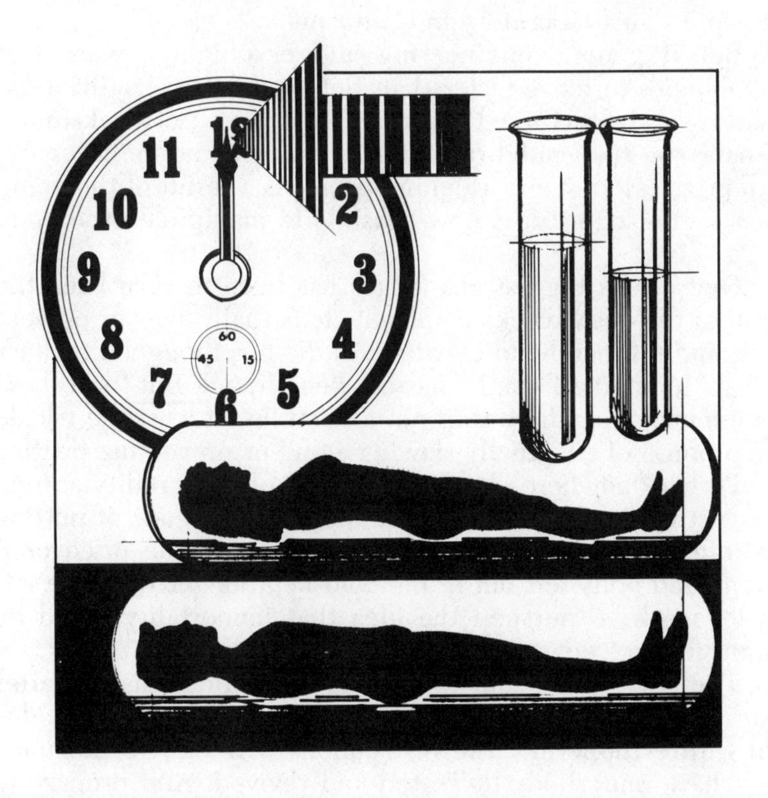

In the movie, *Sleeper*, Woody Allen's character, Miles, experienced the ultimate future shock. Thawed out after two centuries of frozen slumber and unwrapped like a TV dinner, he found himself a stranger in a strange land. Miles had beaten death, but at the cost of having to live in a bizarre world filled with genetically-engineered vegetables the size of Winnebagos, clumsy robot servants, and a president of the United States who had to be cloned from the cells of his own nose.

But Allen's comic vision about life 200 years hence was, like the bespectacled author, somewhat nearsighted. Three years after the movie came out in 1973, Genentech, the first company to commercialize genetic engineering technology, went into business designing new gene products. Then in 1986, just thirteen years after *Sleeper* hit the theaters, the first pig to possess cells with the gene from a cow, was born in Maryland, and a floppy-eared beagle named Miles (after the movie's character), was frozen for one hour and then thawed back to life in a laboratory in California.

But if genetic engineering and cryobiology were two technologies to move faster than fiction, *in vitro* fertilization, where ova are fertilized "in glass" and then put back into a woman's uterus, and brain tissue transplantation have also both progressed at a maddening pace. As a result of these and other technologies, it is now possible to manipulate the body clock.

One of the biggest challenges has been to turn back the hands of the body clock completely, to actually slow or prevent aging and ultimately, to cheat death. So far, though, a technological "fountain of youth" has not been found. But from these new technologies there is, if not a fountain, at least the trickle of a promise of eventually slowing aging or preventing death.

Perhaps nowhere else has the hope of immortality sprung eternal than in the field of cryobiology, the science of putting life into deep freeze. Indeed, ever since people discovered that a dead body left out in the cold kept longer than one left inside, we have pursued the idea that immortality could be sought in the freezer.

The idea goes something like this: Immediately after death, have one's body frozen. Then, at some time in the distant future (hopefully the instructions will survive the journey), have one's body defrosted and revived. And prepare to live again.

In the future, the dream goes, death, aging, and diseases like cancer would be no more serious than a skin rash. In this future time, missing limbs would be replaced. Memories and sexual vigor would be renewed. Age spots and wrinkles would vanish. In this utopia, all that once was would be renewed. That's the dream, anyway. The reality, though, is somewhat different.

The best lesson learned about the revenges of time on the frozen body has come from the study of bodies that have been naturally frozen for a long time and then subsequently thawed. Michael Zimmerman, who is both a professor of Anthropology at the University of Pennsylvania and a pathologist at Hehnemann University, makes a career of doing just that. He studies the remains of ancient animal and human life unearthed from the frozen ground of Alaska. In the 1970s, Zimmerman performed an autopsy on the remains of a woolly mammoth that had been uncovered beneath the frigid Alaskan wasteland. From the radiocarbon dating, Zimmerman discovered that the beast had died some 21,000 years ago and had subsequently been buried under a ton of snow. After some 200 centuries in ice, though, its body was not well preserved; the tissues had deteriorated. Little, save a few muscle tissues, was left to study.

Then, in 1984, Zimmerman and his colleagues discovered the remains of five people in a squashed winter house on a bluff overlooking the Arctic Ocean. The inhabitants had been crushed to death by a wall of sea ice some 470 years ago.

Three of the bodies were but skeletons, but two of the people—both females, were fairly well-preserved. One of the women was determined to be in her mid-twenties when she was killed. Her partially intact body was found under a sleeping platform at the northern end of the house. Within her chest wall, Zimmerman discovered traces of hemoglobin, the chemical in red blood cells that traps oxygen. It had been preserved for over 400 years.

Unlike the younger woman, though, the second, older woman's body (she was determined to be in her forties) was completely intact. Apparently, she had been frozen solid soon after the roof collapsed. The right side of her chest had been crushed and her lungs had collapsed from the impact of the falling roof. But, Zimmerman found her body so well-preserved after 400 years in ice, that he could actually determine what she had eaten that day and the medical state of her body before death. The older woman appeared to have been reproductively active. Her breasts still showed signs of lactation, and there was anatomical evidence that she had delivered a child some two to six months before her death. Moreover, from the state of her beautifully preserved organs, he could tell that at some time during her life, the woman had had pneumonia followed by inflammation of the heart lining, and acute kidney failure. Tragically, she had just recovered from her illness, only to be crushed by her own home.

Today, the diseases this woman suffered from can now be treated, even cured. And if we had the technology to thaw her out and then revive her, she might be walking around today. But four centuries into her future, she is still dead. And there is nothing we can do about that.

Here in this country and at this time, there is a movement afoot to freeze the dead properly, and under controlled conditions. Various organizations, such as the Cryonics Society and the Life Extension Society, have designed do-it-yourself kits for those wishing to spend part of their death within giant thermos bottles tucked into cryogenic warehouses.

Since a body deteriorates minutes after death, amateur cadaver-freezers learn from their kits to be on constant standby. Death may occur at any time and at any place. Because a stranger is unlikely to pick up one's dead husband or wife off the sidewalk and rush them to the nearest frozen foods section of their neighborhood's supermarket, and since most doctors and hospital staff are uneasy about possible law suits, Society members learn to watch their spouses carefully.

A person in ice water will survive for only 20 to 30 minutes before dying. Death comes when the heart stops beating and the core body temperature drops from 98.6 to 77.0 degrees.

If a Society member is "fortunate" to witness their spouse's death in time, they are instructed to quickly carry them to the basement, or to some place where it is cool. They are also told to maintain the circulation of the dead with the careful use of mouth-to-mouth respiration and external heart message, unless, of course, they happen to have a heart-lung machine lying in wait nearby. They must then inject the body with heparin, a blood anti-coagulant, and then cool the body to about 50 degrees with ice packs.

If they get this far, the next step is to perfuse the body for an hour or so with an intravenous solution containing the chemical DMSO or glycerol. These two chemicals keep the cells of the body from forming ice crystals. Ice crystals disrupt cell membranes and destroy tissues.

Once the hard part is done, the body is ready to be frozen. Most instructional packets suggest wrapping the corpse in layers of blankets and dry ice. Dry ice is nothing more than frozen carbon dioxide, but it will keep the body cooled to about minus 150 degrees. But the final stage in the process is to move the body (being careful not to drop it as it will shatter) into a man-sized capsule filled with liquid nitrogen. Liquid nitrogen will freeze the body down to about minus 320 degrees Fahrenheit.

Assuming that one has the money for the procedure and the funds for the lifelong care—about $24,000 by some estimates, not including the yearly "maintenance fees"—and can trust the caretakers not to go out of business or to sell the property to a meat-packing company, and assuming that the equipment still works over the years and that the liquid nitrogen will be replenished every four months, the procedure can still be fraught with problems. One woman was frozen in April of 1967. A few months later, her relatives decided that it was foolish to hope. They removed her frozen body from its thermos bottle, defrosted it, and then buried the thawed woman in the ground.

Another woman died that August. She had even left instructions for her own freezing, and a photograph of herself taken 27 years earlier. Written on the back of the picture were the words: "This is as I wish to be restored." Unfortunately, she died alone and was discovered after three days, too late to do her slightly decomposed looks any good.

Other than leaving a note on one's body saying, "please don't defrost me 'til Christmas 4500 A.D.," there is really nothing one can do to guarantee that he or she will, in the future, be brought back from the dead. The alternative to being frozen after death, then, is to be frozen when one is still alive. But to freeze a living person one would have to overcome some substantial barriers.

Throughout history, there have been stories of people who have, by one accident or another, been frozen only to be revived later. On the 23rd of March 1756, for example, a Swedish peasant staggered out drunk from a tavern and fell asleep in the snow. The next day, his friends found him. Seeing that his body had frozen solid, they decided he was dead, and so carried him home and promptly put him into a coffin. As the afternoon progressed, and the mourners gathered, a doctor came by and decided that the deceased had not been properly examined. He detected no breath or pulse, but then noticed that the man's stomach was a little warm. He then ordered those around him to rub the man's arms and legs while he applied warm cloths to his chest. Miraculously, the man recovered, and the doctor reported the case to the Swedish Academy of Sciences.

A more recent case involved a Chicago woman who, in 1951, was found drunk on a city's sidewalk. She had been out there all night. The temperature that night had reached minus 11 degrees. By the time she was discovered, she was barely breathing and her heartbeat had slowed to a third of what it should have been. Her body temperature was some thirty-five degrees below the normal 98.6 degrees—but she survived.

These freezings were accidental, to be sure. Yet, as of this date, no responsible scientist has purposely frozen a living human body. But, that doesn't mean they aren't trying. At the University of California at Berkeley, for instance, research physiologist Hal Sternberg, and his colleague Paul Segall, are cooling and reviving living dogs with some success.

In 1986, Sternberg successfully "froze" and then revived a three-year-old beagle named Miles. In the experiment, Miles was first given anaesthesia until he was out cold. Then Miles was placed on a bed of ice until his temperature dropped from the normal 101.0 degrees (for a dog) down to 68 degrees. Once his body temperature fell to this level, Sternberg replaced Miles' blood with a clear, synthetic fluid. The synthetic blood, made of balanced salts, glucose sugar, starch, heparin, and buffers (designed to control the accumulation of acids), was used because it wouldn't clot in the cold.

For more than an hour, Miles was legally dead. His body had been cooled to a temperature of about 50 degrees. At that temperature, he did not breathe; his heart did not beat. When the hour was up, Sternberg carefully warmed the dog's body, and put back his blood. The dog survived his adventure. Nine months after the experiment, Miles was still in perfect health.

Miles is Sternberg's success story, of course. Previous dogs weren't so lucky. Some of them suffered epileptic fits after the procedure. Some of the dogs had trouble breathing and later got pneumonia.

But now that they have perfected their technique, Sternberg and Segall want to try the procedure on monkeys, and eventually, on human volunteers. They believe, as do the members of the various cryonics societies, that frozen, people with incurable diseases would eventually be cured in the future. The difference is that people could be frozen alive rather than when they are dead.

The problem with Sternberg's picture is that, so far, his test animals have not been frozen in the technical sense. They have simply been cooled. The difference is important. Cooling the body slows the body's metabolism somewhat. Freezing the body, especially close to absolute zero (−459.69 degrees Fahrenheit) puts a near halt to the motion of the body's molecular reactions.

In Sternberg's cooling technique, there is no circulation of blood or oxygen throughout the body and therefore no oxygen is going to the cells. Without oxygen to stoke their metabolic fires, cells begin to deteriorate. But since they are cooled, they deteriorate very slowly, yet they deteriorate nevertheless. Eventually—it may take months or years—the body will deteriorate to such an extent that it will be nearly impossible to revive it successfully.

But freezing offers protection against the ravages of decomposition. (Leave one steak in the refrigerator for a month and one in the freezer and see which one looks more appetizing.) Decomposition still takes place, but the process is so slowed by the frigid cold that it is almost nonexistent. And since it is likely that cures for cancer, old age, and (hopefully) death will take place very far into the future, it would be advantageous to freeze one's body rather than cool it. But, so far, *that* hasn't been done.

What *has* been done, to a limited extent anyway, has been an attempt to freeze the organs of the body. But, the technical problems and lessons learned are similar to those involved in freezing the whole body.

More than forty years ago, it was easy to freeze any part of the living human body. The real trick was to be able to freeze the material in such a way that it would still be alive when it was defrosted. *That* couldn't be done. Then, in 1948, Audrey Smith and her co-workers at the National Institute of Medical Research in England, discovered that when glycerol was added to fresh frog or rooster semen, the sperm cells within it, once frozen to minus 110 degrees Fahrenheit, could survive after they had been thawed. The thawed sperm cells wiggled their tiny flagella just like fresh sperm cells.

Since Smith's discovery, almost every cell type, from red blood cells to human sperm cells have been frozen and thawed with success. Moreover, the technique has been used on living tissues, as well. Skin, the cornea of the eye, and certain endocrine glands have been preserved for many weeks and defrosted successfully. The snag to the cryopreservation story, however, comes when this technique is used on whole organs.

Audrey Smith discovered this for herself, when in 1957, only nine years after she had stumbled upon the preservative effects of glycerol, she tried to freeze the live hearts of hamsters. In her experiment, she removed the hearts from the anaesthetized animals, slowly perfused the organs with glycerol, and then cooled the hearts down to −110 degrees Fahrenheit. When she defrosted them, though, none of the hearts would beat.

Infectious hepatitis has an incubation period of some 10 to 40 days.

Nowadays, most cryobiologists are trying their luck at freezing and then thawing organs like kidneys and livers freshly removed from anaesthetized animals. But, they are having limited success; the organs that survive the treatment do so only because they were frozen for a few minutes and under the best of conditions. Those organs that were frozen at sub-zero temperatures for longer than an hour, never survived.

The problem in organ cryo-preservation stems mainly from the fact that organs are far bulkier and far more complicated than either tissues or cells. Cells and most tissues of the body can be preserved in a frozen state and then can be thawed successfully because most of the cells are exposed to the environment. Thus they can be cooled and warmed uniformly and completely. Organs, because they are bulkier, have very few of their cells exposed to the environment. They, therefore, freeze faster in one region than another. The uneven freezing damages the entire organ.

For example, a slice of bacon (which has a large surface area and most of its cells exposed to the air) will freeze quickly in the refrigerator and can be thawed out just as fast. But just try to freeze and then thaw out a thick pork chop (which has a small surface area-to-volume ratio as compared to the bacon strip) and see what happens. The outside always freezes and thaws way before the inside does.

By the same token, the large bulk of an organ makes it difficult for the chemical protectants glycerol and DMSO to completely reach all of the cells before freezing occurs. The volume of an organ also makes it difficult for the chemicals to be completely removed before thawing. If the chemicals don't reach all of the cells, or if they are not flushed out of the organ afterwards, the organ dies. But, flooding an organ with these chemicals is not the solution, either. Too high a concentration of either of these two cryoprotectants is toxic to cells. So, when freezing organs, like the liver and the kidney, biologists have to be careful to use just the right amount of chemicals and not a smidgen more.

Second, tissues are made up of only one cell type, all of which freeze at the same time. Organs are made up of a number of different cell types. Each type freezes at a slightly different temperature, requiring cryobiologists to carefully regulate not only the time at which they take to freeze the organ but also the time they take to thaw it out.

The final problem is also the toughest nut to crack. When tissues and cells are frozen, scientists expect to see some ice forming on the outside of the cell membranes. It happens, and no one really cares much about it. But when ice forms within an organ, it is serious. Organs are extremely organized structures. Built up of individual cells, these cells nevertheless form complex structures that connect with other cells. Ice disrupts these structures and so destroys the organ. (All of these problems crop up when people attempt to freeze the dead. So it is unlikely that their organs will completely survive the Great Thaw a thousand years from now.)

The need for frozen organs is obvious to hospitals. If organs could be put into frozen storage indefinitely, then organ transplants would be safer and more successful; doctors would have more time to plan and execute a transplant operation rather than having to rush from donor to recipient before the organ dies. Moreover, organs, placed into frozen storage, could be thawed out in an emergency. (Researchers do it now by popping a frozen organ into a microwave oven.) A patient would not have to go on a heart-lung machine or a kidney dialysis device. The organ would be in the hospital's freezer waiting for them.

But the same technology involved in the cryopreservation of organs will someday make it possible for humans to freeze and then thaw themselves out in the future. It will then be possible to take that long journey to another planet or, perhaps, to another time.

The technology of cryobiology has also smacked up against two newer technologies; those being *in vitro* fertilization and embryo transfer. *In vitro* fertilization allows the marriage of both sperm and egg to occur outside of the womb, usually in a shallow glass dish. Then, the zygote is returned to the uterus. Embryo transfer allows scientists to move an embryo from one uterus to another. Together, they make it possible for, say, a mother to bear her infertile daughter's sons (which actually happened to a woman in South Africa recently). The daughter's eggs were fertilized in a dish and then transferred to her mother's womb.

Cryobiology entered the picture when it was first learned that a woman's ova, or egg cells, could be placed in frozen storage for an extended period of time. The implications were obvious to the medical community. For the first time, one could remove a few of a woman's ova from her ovaries, freeze the eggs, and then implant them back into the uterus at a future date. Or, better yet, if a woman wished to bear twins but could not bear them at the same time, then one of the eggs could be removed and frozen while the first twin grew in the woman's womb. When the first of the twins was born, the second egg (and the twin of the first) could then be defrosted and implanted back into the woman's uterus. Thus, mom would be able to have her cake and eat it too. What gives this latter scenario some importance is that it actually occurred not too long ago, and indeed a little girl did get to watch her twin sister being born.

The bizarre occurrence already has some far-reaching effects for the two twin girls and for the others soon to follow in their footsteps. Identical twins (offspring from the same split egg which thus share the same inherited characteristics) often have a greater probability of dying within a few years of each other than do offspring born from separate eggs.

The Crusi brothers, identical twins, were born in Italy on January 17, 1900. In February 1973, they died. Brother Giuseppe expired 10 days after his twin Lucio. Two years later both John and 66-year-old identical twin Arthur Mowforth, met their deaths. Both British twins died on the same night in May of 1975, and each of them after a heart attack. Brother

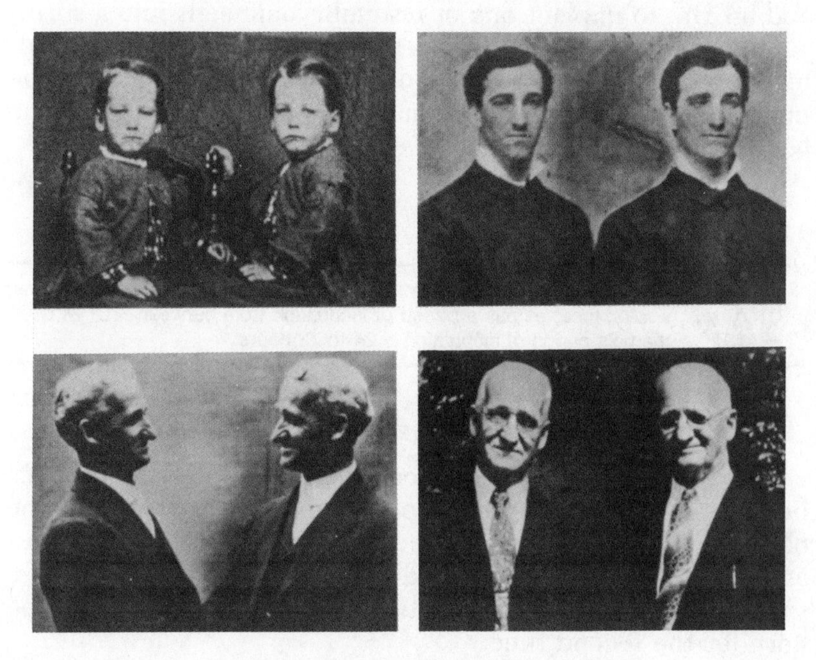

Identical twins at ages 5, 20, 55, and 86.

John expired in a Bristol hospital that night. Arthur died in a hospital in Windsor a few miles away.

To be sure, these are isolated and extraordinary cases. But, even if they only partially represent the norm, the situation makes for weird drama. A genetically identical twin, born years after his identical brother would, for the first time, be able to know the approximate date of his own natural death. He would also know when he would acquire the first signs of a genetic disease, like Huntington's, a severe mental and physical deterioration that occurs in later life. With his older twin brother as his guide, he would be able to peer into his own genetic future.

This is medical ghoulishness, perhaps, but not outlandishly so. What is perhaps truly macabre is the idea that it is now possible to wait for a fertilized egg to divide to four cells, and to remove and freeze three of them. Since each of the four cells has the potential (once each is returned to the uterus) of growing into the same genetically identical child, a parent could present his child with the child's own embryonic cells. It would then be feasible for that individual, when he gets old and infirm, to implant one of his embryonic cells into a surrogate mother's uterus and to be, literally, born again. Then, after his younger twin reaches old age, another cell could be implanted into a surrogate woman's womb, and so on, until the cells are used up.

Chickenpox has an incubation period in the body between two to three weeks; measles has a period of incubation from between 7 to 14 days; the flu's period of incubation is 24 to 72 hours.

Moreover, the procedure could also work for the parents' benefit as the ultimate form of life insurance. If one of the children were to die, say, by accident, the parent could defrost another of the child's cells and implant it into the womb of a surrogate mother. One's genetically identical child would be born for the second time.

Slipping in one's younger genetic double into one's later life is a weak substitute for the real objective: avoiding one's own death. In that, the technology of organ transplantation is on stronger footing. Indeed, ever since Dr. Christiaan Neethling Barnard of South Africa successfully transplanted the young, healthy heart of a twenty-five year old woman, into the chest of fifty-four year old Louis Washkansky on December 3, 1967, the first such operation of its kind, transplanting organs has become nearly routine. Only some twenty years after Barnard's first heart operation, transplant operations have been performed for nearly every organ. For the most part, these transplanted organs have usually been taken from younger and healthier donors (that are close to death, for reasons unrelated to the transplanted organs) and placed into older, more infirm recipients. There can be no doubt that the youth of these organs affords their new human hosts many extra years of life.

But one organ in the human body has been noticeably absent from the list of human tissues up for donation. Unquestionably, the human brain has not been transplanted for both strong ethical and biological reasons; the brain not only harbors one's very identity, it is also exquisitely complex and intricately connected, by a vast array of nerves, to every organ and tissue in the body. The other organs, hearts, livers, and the like, are self-contained units that can be removed from and then reattached to blood vessels. The brain cannot.

This shortcoming has, for some time, left a noticeable bruise in the fight against brain disease. A diseased heart, for example, can now be replaced with a newer model. But how does one go about replacing an old or damaged brain? Actually, one doesn't have to replace the brain. There is a better way.

Just recently, doctors in Mexico successfully treated two patients who had been suffering from the neurological disorder Parkinson's disease by transplanting grafts of tissue from the patients' own adrenal glands into their brains. The adrenal medulla tissue, which is in the interior of the adrenal glands, produces the needed dopamine, a transmitter missing in the brains of these patients, in addition to the major hormones, adrenalin and norepinephrine. The transplants in Mexico followed on the footsteps of an adrenal transplant operation that had been performed on a man in Sweden in 1982.

But adrenal transplants are just the inferior first stage of the treatment for Parkinson's. The next step is to transplant the brain tissue from a fetus into the adult patient. Fetal tissue is perfect. Transplanted into one area of an adult animal's brain, fetal tissue takes on the identity of the cells around it and sends out nerve fibers to other brain areas, making all of the appropriate connections. In mice and rats, and now even in monkeys, transplanted fetal tissue seems to find its way to the right place in the brain and executes the appropriate physiological function.

The fetal cells do this, supposedly, by locking onto particular chemical factors secreted by the injured brain tissue and then sending out their nerve fibers to where the "scent" is strongest. As the fibers travel towards the damaged brain tissue, the fetal cells adopt the identity of the cells that normally connect with this region, even going so far as to manufacture the required transmitter of these cells. Eventually, the fetal cells make the necessary attachments with the cells in this damaged region of the brain by moving in where there is a paucity of old but established connections. Ultimately, the major brain disorders—Alzheimer's, Parkinson's, and perhaps spinal cord injuries—will be treated in this way, using fresh, human fetal tissue.

Ironically, brain grafts are so effective because of the blood-brain barrier. The same cellular wall that prevents many drugs and cells from reaching the brain, also protects the grafted tissue from the cells of the body's immune system. This makes the brain an immunologically isolated region. It is for this reason that brain grafts are not usually rejected.

This advantage was the major reason why researchers Anders Bjorklund and Ulf Stenevi at the University of Lund in Sweden, have been successfully grafting fetal brain tissue in adult brains of rats for more than five years. One of the regions of the brain where they have been grafting brain tissue has been in the hippocampus. The hippocampus, situated roughly in the center of the brain, participates in learning and memory.

First, the Swedish researchers cut into the rats' hippocampus destroying nervous connections. The animals lost their ability to navigate a T maze. Normal rats can easily learn that they must go to alternate sides of a T maze to get food. But these brain-damaged rats never learned; they had lost their short-term memories.

After Bjorklund and Stenevi cut into the rats' brains, they then injected special fetal brain cells, which make the neurotransmitter acetylcholine, into the brains of these damaged adult rats. Miraculously, after two weeks or so, the rats' memories were restored; they could navigate the T maze as well as their normal cousins.

What makes this experiment all the more intriguing is that it bears on the memory loss that occurs with aging and with Alzheimer's disease. Patients with Alzheimer's disease lack brain acetylcholine. Old rats lose their memories and lack acetylcholine, as well. The two researchers are attempting to restore the memories of old rats with this transplantation procedure.

At the same time that the two Swedish researchers were trying to restore rat memories, physicians at Columbia University and at the Mount Sinai School of Medicine were attempting to transplant fetal mice brain tissue into mutant strains of mice that do not make hypothalamic gonadotropin releasing hormone (GnRH). As a consequence of this mutation, males of this mouse strain have immature reproductive organs.

Pneumonia has an incubation period in the body of some one to three days.

But, by transplanting fetal brain tissue into the region of the adult mouse brain responsible for the production of GnRH, the research physicians managed to restore the male mice's sexual vigor. The mice began to make the hormone and their testes got larger.

More exciting, though, is the work being done at the National Institute of Mental Health by William Freed and Richard Jed Wyatt. They are grafting fetal retinas into the superior colliculus—the part of the brain that receives visual information—of blinded adult rats. When animals are blinded, the neural connections between the eye and the superior colliculus deteriorates. Fetal rat retinas, though, can grow and make neural connections when they are grafted onto the superior colliculus. The hope is that the fetal tissue will make connections with the eye itself and so will restore vision to the blind.

But growing human embryos or fetuses as brain tissue farms, raises tremendous moral questions. And so far, the source of embryonic neurons for transplantation will likely be tissue from aborted fetuses. The moral calculus makes it possible that other alternative sources for embryonic tissue will have to be found.

Already, cryobiology and embryo transfer make it possible, though not necessarily ethically palatable, for humans to use their own embryonic tissue by having the first few cells of their own early zygotic divisions placed into frozen storage. Then, when they are old and infirm, they can theoretically defrost one of these cells and have it implanted into a surrogate mother's uterus. The brain tissue resulting from this genetically-identical embryo could be used in the future to replace the old, worn out brain of its owner.

Ultimately, though, if we are to slow aging or alter the biological timepiece of the human body, we must begin by manipulating the body's genetic instructions, its DNA. That task has been left up to one of our newest inventions, gene-splicing, or genetic engineering. This brave new science allows scientists to take apart and recombine the genes of cells.

All the information that makes us up is encoded on our set of 23 chromosome pairs—called the human genome—half of which comes from our father, half from our mother. The chromosomes are made up of DNA and are found in every cell in the body. Each section of DNA on the chromosomes that has a specific function is a gene. A gene usually contains the information to produce a particular protein.

It was, of course, James Watson and Francis Crick, working in England in 1953, who first proposed the "double helix" structure of the DNA molecule. Their efforts not only earned them the Nobel Prize, but also proved to play a major role in later genetic research.

James Watson, along with Francis Crick, proposed the "double-helix" structure of the DNA molecule.

Less than two decades passed before Stanley Cohen of Stanford University and Herbert Boyer of the University of California at San Francisco, found that the newly discovered restriction enzymes, the enzymes bacteria use to carve up foreign DNA, could be used like chemical scissors to selectively "splice" out a specific sequence of DNA from one bacterium and "recombine" it with the DNA of another. The bacterium with this new "recombinant DNA" then could go on to reproduce the genetic code sequence. In effect, Cohen and Boyer showed the scientific community that it was possible to create the kinds of mutations that might take place randomly over millions of years. Only now it could be done in less than an hour.

Armed with this new tool, genetic engineers went off to create such organisms as bacteria that manufacture human insulin and human growth hormone, bacteria that digest oil slicks, and plants that harbor the genes of viruses. The foreign genes make the plants more resistant to diseases and insects.

The technology has also allowed these biologists to create transgenic animals, creatures that have one or two of the genes from other animals inserted in their genetic material. One such animal is a rust-colored boar. The pig, born in 1986 at the Department of Agriculture's laboratory in Maryland, has the growth hormone gene of a cow in its cells. The gene somehow makes the animal leaner than any of its cousins.

Those who use this new tool on bacteria, plants, and animals, are not shy about saying that the technology will be used someday to alter human genetic material, as well. And there would be no dearth of material to alter, either.

Dozens of laboratories have announced the discovery of genes responsible for various inherited diseases by using gene "markers." Gene markers are genes near by on the same chromosome as the errant gene that are used as flags to track the gene in question through successive generations of a family. Scientists find the markers by slicing up chromosomes with enzymes and then look at the resulting pattern of chromosome fragments. Certain fragment patterns are almost always inherited with the genetic disease. Such markers have been used to locate the genes for Huntington's disease, muscular dystrophy, cystic fibrosis, and polycystic kidney. Most recently, the gene responsible for manic depression was located through its markers, and the gene for Alzheimer's disease was found, too.

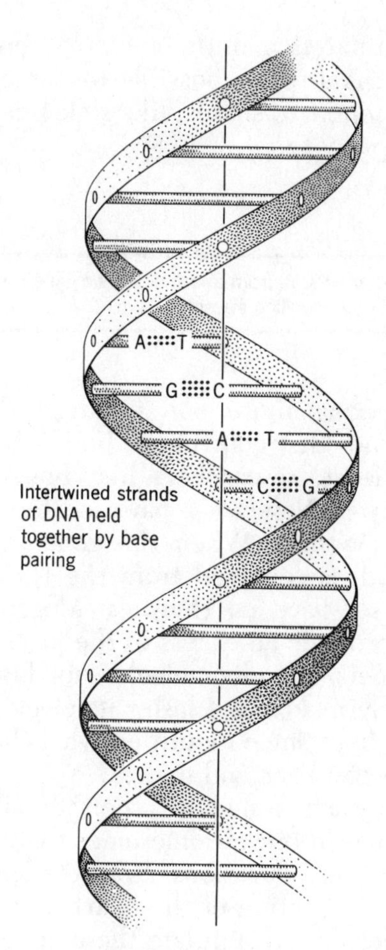

The DNA molecule.

Armed with that bit of information, scientists can identify who carries the gene in families prone to a particular genetic ailment. All they need are blood samples from a known carrier of the disease gene, and one from a known noncarrier. But, it is only a matter of time when scientists will be able to isolate the disease genes themselves and remove them, or will be able to insert normal ones.

But why stop there? According to the latest estimates, there are between 100,000 and 300,000 genes that make up the human body. So far, though, only a few hundred of them have been fully isolated and their entire genetic sequence worked out. But it is already possible to detect the genes of some of the 3000 genetic diseases, like sickle-cell anemia, that many humans are privy to.

Mumps are infectious from about seven days before the first symptoms appear until the swelling subsides.

Somewhere among those one hundred thousand or so genes there are genes that control the body's biological clock. In fruit flies, at least, researchers like Ronald Konopka of Clarkson University in New York, have located the genes that control circadian rhythms. When the gene, which Konopka calls *per* for period, is removed from the flies, for example, their body processes have no rhythms. Moreover, when the flies are made to produce an excess of the protein that the *per* gene codes for, their circadian clocks run fast. In fact, the more protein the fly makes, the faster its clock runs. Konopka and his colleagues have since identified what they think might be the *per* gene in chickens and mice.

If such a gene exists in humans, then it is likely that there might be other genes in our chromosomes that would code for other body rhythms and durations, from the speed of a nerve signal and the steady beating of the heart, to the onset of puberty. But how does one manipulate these genes?

Scientists have found a number of ingenious tricks to get an organism to express the genes of another. One method simply involves injecting the genes of another creature directly into a fertilized egg with a needle. This is how the Department of Agriculture created their transgenic pig. But piercing cell membranes with a needle kills three quarters of the eggs. In the four years that they have worked on the project, the government researchers have had to inject more than 8,000 fertilized eggs before they could produce just 43 transgenic pigs.

A far more delicate method is to have a virus do the job itself. Viruses are essentially microscopic packages of genetic material surrounded by a coat of protein. They enter cells easily and once inside, many of them splice their own genetic

A chromosome.

material into the genetic material of their hosts. Unable to identify foreign DNA from their own, cells end up making fresh viruses as well as their own cellular products.

At the Massachusetts Institute of Technology, Richard Mulligan and his colleagues have been using retroviruses to transfer genes into mouse cells. Retroviruses are unusual because, unlike most other viruses, they do not kill cells. Instead they enter when the cells are dividing and insert their viral genes into the cells' chromosomes. Mulligan's goal is to use these viruses to treat patients with rare inherited diseases. The viruses would serve as vehicles for carrying within them certain therapeutically valuable genes into a patient's defective cells.

But Mulligan did not want the virus to leave the cells and infect other cells once it had inserted its genes into the first cells. So, in 1983, he and Nobel Prize recipient David Baltimore, designed this virus so that the gene that codes for its protein coat was missing. Without the gene to make its protective coat, the virus can never leave the cell. Now Mulligan's retrovirus can get into a cell and insert its genes into the cell's chromosomes. But the virus cannot be transmitted any further because it lacks the gene to make its coat.

Since Mulligan redesigned the virus for his purposes, he has used the virus ala foreign gene to infect mouse bone marrow cells. Bone marrow cells contain among them stem cells, or precursor marrow cells, that divide indefinitely. Stem cells are the mothers of blood cells. Once a virus injects its foreign gene into a stem cell, every type of blood cell in the body will carry that gene in their chromosomes. Eventually, blood diseases such as sickle cell anemia and thalassemia will be cured by Milligan's method of viral gene transplantation. Eventually, too, Mulligan's tool could be used to inject healthy human genes into the greatest mother stem cell of them all, the fertilized egg.

By 1999, the entire human genome will have been charted and mapped, this according to Nobel Prize winners Walter Gilbert and James Watson. It would be a vast undertaking, to be sure. Written out using its four-letter alphabet of A,G,T, and C (standing for the bases that make up the DNA chain), the entire human DNA sequence would take up the space of a thousand New York City telephone books. But when it is done, it would create a Rosetta stone, a vast library of information about *Homo sapiens*.

Once sequenced, the human genome could be compared with the sequences of other animals to see where it differs. Or, the genome from a healthy individual could be compared to the genome of one suffering from cancer.

Eventually, the day will come when human enzymes are redesigned to work faster or more efficiently, or when genes are manipulated to speed up or slow down the body clock. Some day, the steady drumbeat of time on the human body will be controlled or even eradicated. Then, aging will be halted, death will be postponed.

When that day comes, we shall have finally conquered Time.

APPENDIX

APPENDIX

The Life of Enzymes (The Body in Microseconds)

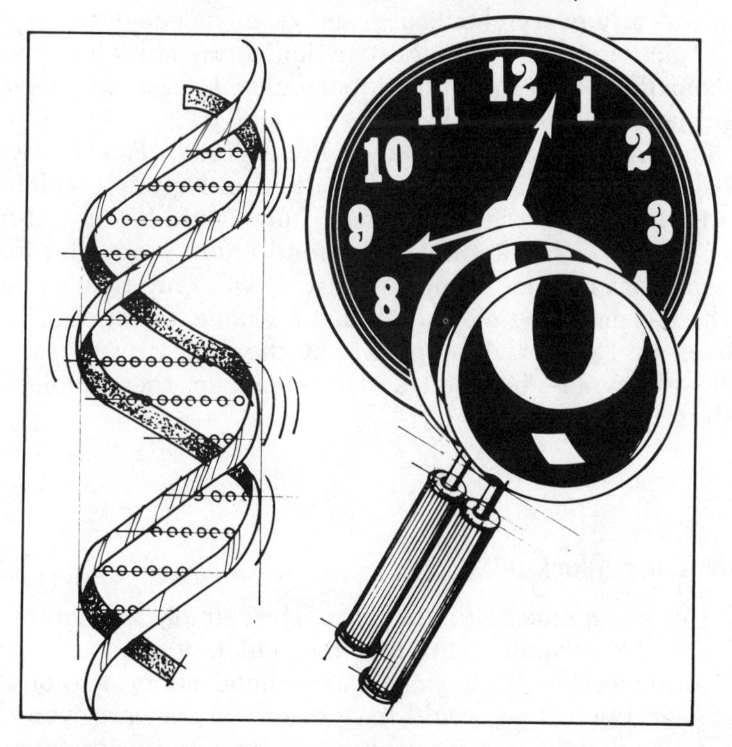

Underlying everything in the book is the life of the enzymes, many of which perform their jobs within a few microseconds. So for you more inclined to delve into learning about their complex behavior, this is their story:

A cockroach knows a lot about motion. Two hundred and fifty million years of evolution and city living has made the roach an expert on the subject. Just try to catch one. It flits away faster than a stamping foot.

If roaches moved as slowly as the human leg hovering above them, there would doubtless be many more squashed insect carcasses around. But, they don't.

There is a marvelously adaptive quality to evolution. Survival is often awarded to the fastest. A fly may have a response time of 65 milliseconds, but its predator, the praying mantis, is five milliseconds quicker.

Evolution's rules apply to enzymes, as well. Biochemists have noticed a fascinating trend for these catalytic molecules to become, if not faster, then more efficient.

There are about 3,000 different enzymes in a cell. They control nearly every chemical reaction, influence every aspect of cell activity, from the construction of membrane parts, to the hoarding of energy in food molecules. Life, as we know it, would not exist without them.

The first ancients to uncover these protein catalysts were probably the Greeks who accidentally (but happily!) stumbled onto the discovery that their grape juice could be turned into wine. The process was only understood 2,000 years later to be the work of enzymes found in yeast. It was only fitting, then, for the German physiologist, Wilhelm Kuhne, to raise his glass of beer in 1876, and propose that the Greek word for "in yeast"—enzyme—be used as the name for these biological catalysts.

How They Work

There is no magic to enzymes. They simply accelerate reactions, like turning sugar into molecules of carbon dioxide and water—often by as much as a hundred or a thousand times—reactions that would have occurred anyway given that one had a month or even a year to wait around for them to happen. What's more, enzymes do this without actually being used up in the process. This gives them tremendous staying power, for they can be used over and over again. Molecular specialists, they are tailored to catalyze only one type of reaction. This makes them highly specific, and far easier to control.

Carbonic anhydrase, for example, is an enzyme that speeds up the formation of carbonic acid from carbon dioxide

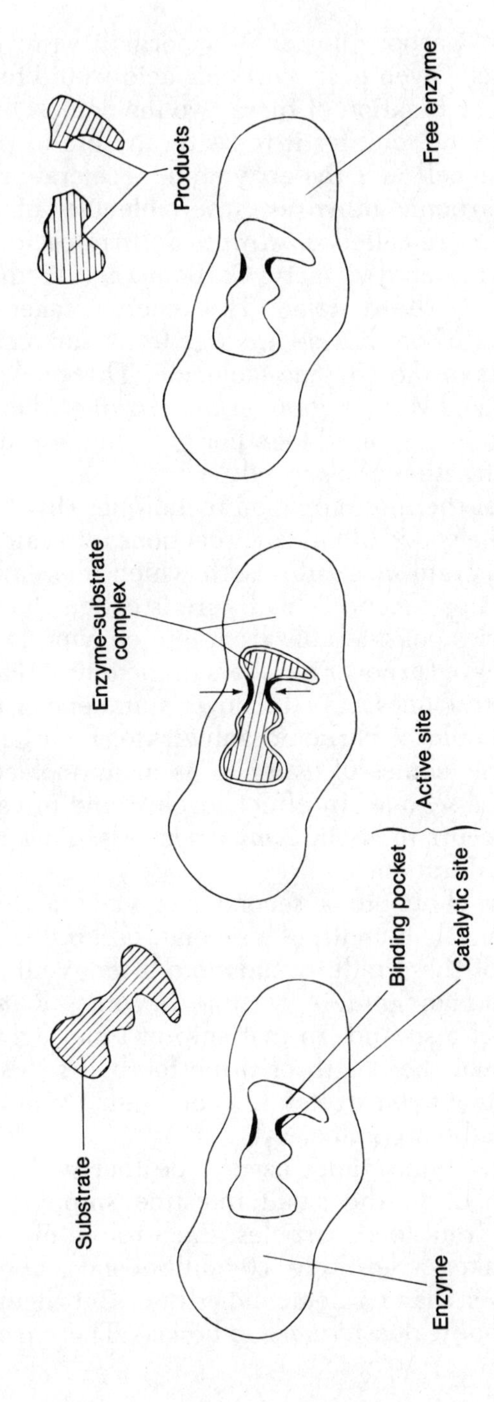

Enzymes are catalysts that make reactions go faster. They turn substrates into products.

and water. Carbon dioxide is a normal waste product of the body's cells. Given time, carbonic acid would form all by itself from the combination of these two molecules, but the reaction would take too long for it to serve any useful purpose for the cell. So the cell uses the enzyme to accelerate the reaction.

For carbonic anhydrase, the molecules of water and carbon dioxide are called *substrates*. Substrates are the molecules on which the enzyme acts. Carbonic acid is the *product*, the end result of the reaction. The reaction takes place because water and carbon dioxide are able to fit into a catalytic "pocket" of sorts in the enzyme molecule. This pocket is called the *active site* and is the region on the enzyme where carbon dioxide and water are fused together to form the product, carbonic acid, in a fraction of a second.

One of the most common techniques that biochemists use to judge the speed of enzyme reactions is to calculate the number of substrate molecules with which one enzyme molecule will react in a second. This figure is called the turnover number. It varies considerably from one enzyme to another. Most enzymes have turnover numbers in the tens or hundreds. A few of the fastest ones have turnover numbers in the thousands. Each molecule of carbonic anhydrase, for example, can fuse 600,000 molecules of water to as many molecules of carbon dioxide in a second. In effect, each round of catalysis for this enzyme occurs in about 2 *micro*seconds. That is an unbelievably short duration.

Finely chopped, a second can yield a thousand bits of time, each a thousandth of a second, or a *millisecond* long. But dice one of these milliseconds into a thousand other bits, and each lilliputian granule of time becomes a *micro*second—a millionth of a second. In that snip of time, light just manages to bolt down the length of three football fields. Or, to put it another way, if you stretch a second into a year, a microsecond would be a beer commercial.

Most enzymes don't have to be that swift. Chymotrypsin putters about in the small intestine snipping small protein molecules from food particles. Each round of catalysis for this enzyme takes a leisurely 10 milliseconds, about a thousand times slower than carbonic anhydrase. But digestion, while vital to the body, doesn't *have* to be fast. There may, in fact, be a distinct evolutionary advantage to having this reaction work at the pace it does. The activity of chymotrypsin may match the speed at which the cells that line the digestive tract can pull in food molecules from inside the gut.

The DNA Enzyme

If digestive enzymes like chymotrypsin can afford to take their time, the workhorse catalyst known as DNA polymerase One, cannot. DNA polymerase has the job of correcting the mistakes in the DNA double helix.

Everything we are is determined by the chemical information inscribed in the DNA molecule. The DNA molecule is made of two strands of genetic material wrapped around each other in spiraling mirror images. The information is encoded into four chemicals or bases known as adenine, cytosine, guanine, and thymine, abbreviated as A, C, G, and T. The combination of these four bases, always in matching pairs—adenine with thymine, guanine with cytosine, in billions of diverse ways, makes the genetic code different for every organism on the planet.

Most of the time the genetic code is error-free. But, sometimes, mistakes happen. The mistakes occur when one base molecule, such as cytosine, on one strand of the DNA double spiral, is damaged. The damaged base no longer binds with its base partner on the other DNA strand. To DNA polymerase, that mismatched base pair sticks out like a warning flag on the DNA molecule and the enzyme easily locates the disfigured base pair.

DNA polymerase One removes the damaged base molecule and inserts a correct one in its place by using the bases on the opposite strand of the DNA double helix as a template. Step-by-careful-step, it "reads" the base on the opposite template strand and sees whether the hydrogen bonds match up with the base opposite it. If it does not, the mismatched base molecule is removed and the enzyme attaches the correct one to the DNA strand. Moreover, DNA polymerase checks its steps twice. First it checks to see that the base it is attaching

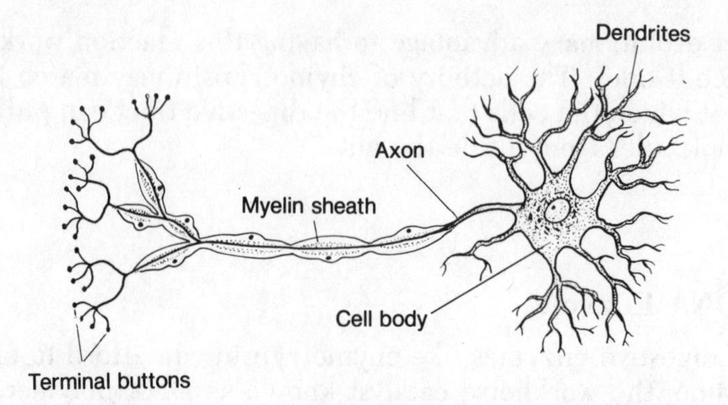

A nerve cell, or neuron. Nervous signals travel down its length from the dendrites to the axon. They trigger the release of chemical transmitters at the buttons or synapses.

interlocks with its sister across the way. Then it checks it again when it has finished its job. Only then does it move on. In fact, it won't budge until both inspections are made.

DNA polymerase completes about ten bases every second. As enzymes go, that is incredibly slow. Apparently, this enzyme trades off speed for accuracy and, DNA polymerase has to be meticulous. One slip up, one incorrect base, and the cell has a major identity crisis—a mutation, in biological parlance. So DNA polymerase *is* accurate, painstakingly accurate. Biochemists have calculated that this enzyme goofs up only about once for every 10 billion bases that it constructs.

An enzyme like DNA polymerase works nicely in places where quality control is paramount. Certainly cells would not have it any other way. But there are other places in the body where an enzyme with lightning speed is what's needed—fast. Meticulous accuracy be damned. Carbonic anhydrase is one such enzyme. Choline esterase is the other.

Nervous enzymes

Choline esterase holes up in a place called the *synapse.* A synapse is the junction where nerve cells interact. Actually, this enzyme can be found only in select synapses in certain areas of the brain and in nerve cells that control the activities of the heart, sweat glands, and intestine. Perhaps more importantly, choline esterase is stuffed into the junctions between the nerves and skeletal muscles.

A nerve communicates with its neighbor by sending an electrical signal down its length until the signal smashes into a synapse and stops. At the synapse is a minute gap or space separating the two nerve cells, called the *synaptic cleft*. In order for the sending nerve to get its message to the receiving nerve, it must secrete a chemical, or transmitter, into the synaptic gap. This chemical secretion product is called *acetylcholine*.

When a nervous signal arrives at the synapse, hordes of acetylcholine molecules wash across this gap, eventually attaching to receptors on the other shore, triggering the other cell into action. Muscles contract. A limb twitches.

If this were all there were to it, nervous communication would be ineffective. Nerves send messages by a system of Morse code but use only dots. The more important the message, the more frequent the dots. And each dot—each nervous signal—triggers its own release of acetylcholine. That is why the field of transmitters must be first cleared away before another signal comes down the pike. Otherwise, the message gets garbled. For nerves that must at times transmit 500 of these nervous signals every second, it means the hordes of acetylcholine transmitters must be destroyed within a fraction of a millisecond.

Choline esterase enzymes are designed to do just this. The distant synaptic shore is packed with these enzymes. Like seagulls after newly hatched turtles, these catalysts tear into acetylcholine molecules with merciless efficiency. Biochemists have clocked them munching on transmitters at the rate of 25,000 molecules per second. Put another way, each hungry esterase rips apart an acetylcholine molecule in about 40 microseconds.

Oddly, choline esterase can be defeated. The tropical African climbing plant, *Physostigma*, for example, grows a dark brown fruit called the Calabar bean. The bean contains the chemical, physostigmine, that binds to the enzyme making it useless and causing sudden respiratory paralysis.

Certain African tribes used the poison from the Calabar bean for their "trials by ordeal" to determine whom among them was a witch. Presumably, the ones who died were the innocent ones.

Physostigmine is not the only choline esterase inhibitor. There are a few others. Perhaps some of the most effective are organic fluorophosphates such as DIPF. Each is designed to stuff up the active site of the esterase molecule. Without an active site, the enzyme doesn't work. One or two of them have been inserted into bug sprays. A couple of others have been made into nerve gas.

Perfect Catalysts

Choline esterase is known, in the biochemical community, as a perfect enzyme. To win this prestigious title requires it to meet just one criterion: An enzyme has to be so fast at tearing off the molecular heads of its victims that the only delay in its activity is in the time it takes for the substrate

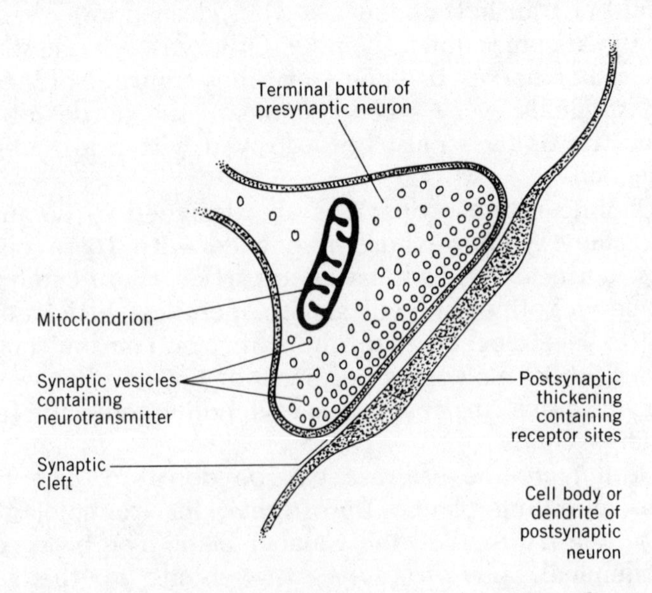

Acetylcholine, and its enzyme choline esterase, work at the synapse.

molecules to meander to it. Those enzymes that are so efficient that they have to wait for their meals, are said to have obtained kinetic perfection. They are as perfectly designed as any enzyme could be. Choline esterase falls into this category, as does carbonic anhydrase and several others. They cannot get any faster.

Well, not quite. Evolution has stumbled upon an interesting device to counter the effects of this inefficient substrate meandering—a problem that choline esterase still has to face, simply by eliminating it altogether. Through the millennia, a certain number of these enzymes have clustered together into giant complexes. Like a bucket brigade, one enzyme simply hands over its product—now substrate—to the next enzyme nearby. Then that enzyme hands its product to the next enzyme, and so on until the final product is made. In this system, there is a minimum amount of waiting and little delay. The substrates and the products are kept together within the enzyme cluster.

One of the more interesting enzyme assemblies that has adopted this shortcut is the pyruvate dehydrogenase complex, PDH. PDH, which is made up of three different kinds of enzymes, lies in each of our cell's mitochondria. Mitochondria are tiny bacteria-like powerhouses that provide most of the energy for the entire cell. PDH is vitally important to mitochondria because it helps to keep the food molecules tagged for burning within the confines of these structures.

One of the major fuels used by the cells of the body is glucose sugar. There is an ancient ritual in the burning of glucose, one that appears to have been passed down from cellular ancestors billions of years ago. Indeed, yeast cells still use much of the process, called *fermentation,* to this day in their quest to transform sugar into alcohol.

But in the cells of our body, glucose is broken down, step by enzymatic step, to the last products on the enzymatic conveyer belt, molecules of pyruvate. The process, called *glycolysis,* is fairly slow—it takes time for substrates to drift over to their enzymes for conversion—and inefficient, sequestering only a meager amount of energy. In effect, cells that use only glycolysis eat just the mayonnaise on the bread, and leave the rest of the sandwich on the plate. Very inefficient, indeed.

The cell's mitochondria have solved this problem. They are designed to squeeze the last drop of energy out of pyruvate. The mitochondria "burn" this molecule left over from the glycolytic pathway to carbon dioxide and water and produce *lots* of energy for the rest of the cell to use.

But to do this, pyruvate first has to be trapped within the mitochondrion. PDH does just this, converting pyruvate to a molecule called acetyl co-enzyme A.

The procedure is rather complex. PDH has to snatch hold of the pyruvate molecule, pop off an atom, then attach the atom to oxygen to make carbon dioxide. At the same time it is doing that, it also has to grab onto a co-enzyme A molecule, then attach the altered pyruvate, now called an acetyl molecule, to a molecule of co-enzyme A to make the final product molecule of acetyl co-enzyme A. It sounds complicated (and it is). But, if the reaction were done by three separate enzymes floating around in the mitochondrion, the process might take several milliseconds. Instead, the catalytic reaction is far more rapid; the three enzymes that perform this feat are clustered into one efficient complex.

Evolution created a supremely efficient molecular body. But this body has its Achilles heel. PDH requires the vitamin thiamine to function properly. Without it, PDH runs very slowly or not at all.

The lack of thiamine in the diet leads to what the inhabitants of the island of Java once called, Beri-beri, a word meaning, sheep. "I believe those, whom this same disease attacks, with their knees shaking and the legs raised up, walk like sheep," Dutch physician Jacobus Bonitus wrote in 1630. "It is a kind of paralysis, or rather Tremor: for it penetrates the motion and sensation of the hands and feet. . . ." For those without this necessary vitamin, the heart soon weakens, the nerves deteriorate.

In the years since Jacobus Bonitus discovered what happens when our biochemistry is interrupted, humans in their scientific endeavors, have found that the evolution of life is no longer confined to things we can see, but occurs even in the smallest of biological life—cells, enzymes, fatty membrane molecules, as well.

Today, molecular biologists are discovering that these tiny

molecules have a life of their own, far more active than anything we could have imagined. This fact, above all, is paramount. For without their rapid-fire actions preparing our nerves, our muscles, our very thoughts for action, we would have died out long ago, devoured by life far more efficient than ourselves.

REFERENCES

CHAPTER ONE

Argyle, M. and M. Cook. *Gaze and Mutual Gaze*. Cambridge University Press. London. 1976.

Black, D. W. 1984. Laughter. JAMA. *252:* 2995–2998.

Critical Periods. (J. P. Scott. Ed.) Dowden Hutchinson & Ross, Inc. Pennsylvania. 1978.

Davidson, J. M. 1981. The orgasmic connection. Psych. Today. *15:* 91.

Elton, L. R. B. and H. Messel. *Time and Man.* Pergamon Press, New York. 1978.

Hall, B. and J. J. Niles. *One Man's War.* Arno Press. New York. 1980.

Hass, H. *The Human Animal.* G. P. Putnam's Sons. New York. 1970.

Hendrickson, R. A. *The Rise and Fall of Alexander Hamilton.* Van Nostrand Reinhold, New York. 1981.

Gallup, G. G. and S. D. Suarez. 1983. Optimal reproductive strategies for bipedalism. J. Hum. Evol. *12:* 193–196.

Lewis, K. R. and H. Plotkin. 1983. The lightning calculator. The Sciences. *23:* 45.

Mason, H. M. *High Flew The Falcons.* J. B. Lippincott Co. New York. 1965.

Menaker, W. and A. Menaker. 1959. Lunar periodicity in human reproduction: a likely unit of biological time. Am. J. Obst. Gynec. *77:* 905–914.

Morris, D. *Intimate Behavior.* Random House. New York. 1971.

Palmer, J. D., J. R. Udry, and N. M. Morris. 1982. Diurnal and weekly, but no lunar rhythms in human copulation. Human Biol. *54:* 111–121.

Saks, S. *The Craft of Comedy Writing.* Writer's Digest Books, Cincinnati. 1985.

Sternberg, R. J. 1986. Inside intelligence. American Scientist. (March-April): 137–143.

Symons, D. *The Evolution of Human Sexuality.* Oxford University Press. New York. 1979.

The Development of Expressive Behavior. (G. Zivin, Ed.) Academic Press. New York. 1985.

The Voices of Time. (J. T. Fraser, Ed.) George Braziller Publ. New York. 1966.

Timing and Time Perception. (J. Gibbon and L. Allan, Eds.) Annals of the New York Academy of Sciences (Volume 423). The New York Academy of Sciences. New York. 1984.

Western Sexuality. (P. Aries, and A. Bejin., Eds.) Basil Blackwell Ltd. 1985.

Zuckerman, M. 1983. Sexual arousal in the human: Love, chemistry or conditioning? In: *Physiological Correlates of Human Behavior.* Vol. 1. Academic Press, London. (Gale, A. and J. A. Edwards, Eds.) pp. 299–325.

CHAPTER TWO

Birch, C. A. 1959. Sneezing. The Practitioner. *182:* 122–124.

Brouillette, R. T., B. T. Thach, Y. K. Abu-Osba, and S. L. Wilson. 1980. Hiccups in infants: characteristics and effects on ventilation. J. Pediatr. *96:* 219–225.

Co, S. 1979. Intractable sneezing: case report and literature review. Arch. Neurol. *36:* 111–112.

Davis, J. N. 1970. An experimenal study of hiccup.

Doane, M. G. 1984. Turnover and drainage of tears. Ann. Opthalmol. *16:* 111–114.

Hall, A. 1945. The origin and purposes of blinking. Br. J. Ophthalmol. *29:* 445–467.

Jankovic, J., W. E. Havins, R. B. Wilkins. 1982. Blinking and blepharospasm. JAMA. *248:* 3160–3163.

Karson, C. N., K. F. Berman, E. F. Dollelly, W. B. Mendelson, J. E. Kleinman, and R. J. Wyatt. 1981. Speaking, thinking, and blinking. Psychiatry Res. *5:* 243–246.

Kavka, S. J. 1983. The sneeze—blissful or baneful? JAMA. *249:* 2304–5.

Kennard, D. W. and G. H. Glaser. 1964. An analysis of eyelid movements. J. Nerv. Ment. Dis. *139:* 31–48.

Nathan, M. D., R. T. Leshner, and A. P. Keller. 1980. Intractable hiccups. The Laryngoscope. *90:* 1612–1618.

Ross, B. B., R. Gramiak, and H. Rahn. 1955. Physical dynamics of the cough mechanism. J. Appl. Physiol. *8:* 264–268.

Stern, J. A., L. C. Walrath, and R. Goldstein. 1984. The endogenous eyeblink. Psychophysiology. *21:* 22–28.

Stromberg, B. V. 1979. The hiccup. Eye, Ear, Nose, Throat J. *58:* 354–357.

Stromberg, B. V. 1975. The sneeze. Eye, Ear, Nose, and Throat J. *54:* 450–453.

von Cramon, D. and U. Schuri. 1980. Blink frequency and speech motor activity. Neuropsychologia. *18:* 603–606.

Wagner, M. S. and J. S. Stapczynski. 1982. Persistent hiccups. Ann. Emerg. Med. *11:* 24/47–26/49.

CHAPTER THREE

Biologic Basis of Wound Healing. (Lewis Menaker, Ed.) Harper and Row, Publ. New York. 1975.

Blum, H. F. *Photodynamic Action and Diseases Caused by Light.* Hafner Publ. Co. New York. 1964.

Bordicks, K. J. *Patterns of Shock: Implications for Nursing Care.* Macmillan Publ. Co., New York. 1980.
Guyton, A. C. *Textbook of Medical Physiology.* Fifth Edition. W. B. Saunders Company. Philadelphia. 1976.
Massie, R. K. *Nicholas and Alexandra.* Atheneum. New York. 1968.
The New York Times. March 31, 1981.
O'Boyle, C. M., D. K. Davis, B. A. Russo, T. J. Kraf. *Emergency Care: The First 24 Hours.* Appleton-Century-Crofts. Norwalk, Connecticut. 1985.
Shock. (Nursing Now Series). Springhouse Corporation. Pennsylvania. 1984.
The Washington Post. March 31, 1981.

CHAPTER FOUR

Asimov, I. *A Short History of Biology.* The Natural History Press, New York. 1964.
Barnes, D. M. 1986. Steroids may influence changes in mood. Science. *232:* 1344–1345.
Bayer, M. J., B. H. Rumack, and L. A. Wanke. *Toxicologic Emergencies.* Robert J. Brady, Co. Maryland. 1984.
DiPalma, J. R. *Basic Pharmacology in Medicine.* McGraw Hill. New York. 1976.
Fabro, S. and S. M. Sieber. 1969. Caffeine and nicotine penetrate the pre-implantation blastocyst. Nature. *223:* 410–411.
Iversen, S. D. and L. L. Iversen. *Behavioral Pharmacology.* Oxford University Press. 1981.
Jensen, L. B. *Poisoning Misadventures.* Charles C. Thomas, Publ. Illinois. 1970.
Julien, R. M. *A Primer of Drug Action.* W. H. Freeman and Company. New York. 1985.
Kandel, E. R. and J. H. Schwartz. *Principles of Neural Science.* Elsevier/North-Holland. New York. 1981.
Manvell, R. and H. Fraenkel. *Himmler.* G. P. Putnam's Sons. New York. 1965.
Sershen, H. and A. Lajtha. 1979. Cerebral uptake of nicotine and of amino acids. J. Neurosci. Res. *4:* 85–91.
Sershen, H., M. E. A. Reith, M. Banay-Schwartz, A. Lajtha. 1982. Effects of prenatal administration of nicotine on amino acid pools, protein metabolism, and nicotine binding in the brain. Neurochem. Res. *7:* 1515–1522.
Silverman, H. M. and G. I. Simon. *The Pill Book.* Bantam. New York. 1980.
de Wied, D. 1980. Hormonal influences on motivation, learning, memory, and psychosis. In: *Neuroendocrinology.* (D. T. Krieger and J. C. Hughes, Eds.) Sinauer Assoc. Sunderland, Mass. pp.194–204.
Weider, B. and D. Hapgood. *The Murder of Napoleon.* Congdon & Lattes, Inc. New York. 1982.
Winter, P. M. and J. N. Miller. 1985. Anesthesiology. Sci. Am. *252:* 124–131.

CHAPTER FIVE

Abrams, I. S. 1986. Beyond night and day. Space World. (December) pp. 12–13.

Aschoff, J. 1980. The circadian system in man. In: *Neuroendocrinology.* (D. T. Krieger and J. C. Hughes, Eds.) Sinauer Assoc. Sunderland, Mass. pp. 215–222.

Barnes, P., G. Fitzgerald, M. Brown, C. Dollery. 1980. Nocturnal asthma and changes in circulating epinephrine, histamine, and cortisol. N. Engl. J. Med. *303:* 263–267.

Biological Rhythms and Medicine. (A. Reinberg and M. H. Smolensky, Eds.), Springer-Verlag, New York. 1983.

Biological Rhythms, Sleep, and Performance. (W. B. Webb, Ed.) John Wiley and Sons, New York. 1982.

Borbely, A. *Secrets of Sleep.* Basic Books, Inc. New York. 1986.

Cellular Pacemakers. Vol. 2. (D. O. Carpenter, Ed.) John Wiley and Sons, New York, 1982.

"Circadian variation in ozone tolerance." 1987. Science News. *131*(11): 169.

Circadian Rhythms in the Central Nervous System. (Redfern, P. H., I. C. Campbell, J. A. Davies, and K. F. Martin, Eds.) VCH, London. 1985.

Clark, T. J. H. and M. R. Hetzel. 1977. Diurnal variation of asthma. Br. J. Dis. Chest. *71:* 87–92.

Czeisler, C. A., J. S. Allan, S. H. Strogatz, J. M. Ronda, R. Sanchez, C. D. Rios, W. O. Freitag, G. S. Richardson, and R. E. Kronauer. 1986. Bright light resets the human circadian pacemaker independent of the timing of the sleep-wake cycle. Science. *233:* 667–671.

"Daily rhythms running like clockwork." Science News. *130*(9): 136.

"Deadly blooms and curious clocks." 1987. Science News. *131*(8): 122.

Gautherie, M., C. Gros. 1977. Circadian rhythm alteration of skin temperature in breast cancer. Chronobiologia. *4:* 1–17.

Halberg, F. 1980. Implications of biologic rhythms for clinical practice. In: *Neuroendocrinology.* (D. T. Krieger and J. C. Hughes. Eds.). Sinauer Assoc. Sunderland, Mass. pp. 109–122.

Herbert, W. 1982. Punching the biological timeclock. Science News. *122*(5):69.

Hilts, P. 1980. The clock within. Science 80. *1:* 61–67.

Hogarth, P. J. *Biology of Reproduction.* John Wiley and Sons. New York. 1978.

Kolata, G. 1985. Finding biological clocks in fetuses. Science. *230:* 929–930.

Kolata, G. 1985. Genes and biological clocks. Science. *230:* 1151–1152.

Kolata, G. 1986. Heart attacks at 9:00 a.m. Science. *233:* 417–418.

Lagercrantz, H. and T. A. Slotkin. 1986. The "stress" of being born. Sci. Am. *254*(4): 100–107.

Lindsley, J. G. 1983. Sleep patterns and functions. In: *Physiological Correlates of Human Behavior*. Vol. 1. (Gale, A. and J. A. Edwards, Eds.) Academic Press, London. pp.105–139.

Minors, D. S. and J. M. Waterhouse. *Circadian Rhythms and the Human.* Wright-PSG, London. 1981.

Monk, T. H. and J. C. Gillin. 1984. Circadian lability and shift work intolerance. TINS. (December) pp.459–460.

Moore, R. Y. 1982. The suprachiasmatic nucleus and the organization of a circadian system. TINS. (November). pp. 404–407.

Olmert, M. 1984. Points of origin. Smithsonian. *15:* 38–40.

Pekkanen, J. 1982. Why do we sleep? Science 82. (June). pp. 86.

Thompson, M. J. and D. W. Harsha. 1984. Our rhythms still follow ths African sun. Psychology Today. (January). pp.52–54.

Vertebrate Circadian Systems. (Aschoff, J., S. Daan, G. A. Groos, Eds.), Springer-Verlag, New York. 1982.

Zucker, I. 1983. Vertebrate Circadian Systems. (Review). Science. *220:* 854–855.

CHAPTER SIX

von Baumgarten, R., A. Benson, A. Berthoz, Th. Brandt, U. Brand, W. Bruzek, J. Kass, Th. Probst, H. Scherer, T. Vieville, H. Vogel, J. Wetzig. 1984. Effects of rectilinear acceleration and optokinetic and caloric stimulations in space. Science. *225:* 208–212.

Cogoli, A., A. Tschopp, and P. Fuchs-Bislin. 1984. Cell sensitivity to gravity. Science. *225:* 229–230.

Dorr, L. 1987. When the doctor is 200 miles away. Space World. *X-3-279:* 33–36.

Environmental Physiology: aging, heat and altitude. (S. M. Horvath and M. K. Yousef, Eds.) Elsevier/North-Holland. New York. 1981.

Frisancho, A. R. *Human Adaptation.* The C. V. Mosby Co. St. Louis, MO. 1979.

High Altitude Physiology. (J. B. West, Ed.) Hutchinson Ross Publ. Co. Pennsylvania. 1981.

Kirsch, K. A., L. Rocker, O. H. Gauer, R. Krause, C. Leach, H. J. Wicke, R. Landry. 1984. Venous pressure in man during weightlessness. Science. *225:* 218–219.

Leach, C. S. and P. C. Johnson. 1984. Influence of spaceflight on erythrokinetics in man. Science. *225:* 216–218.

Quadens, O., and H. Green. 1984. Eye movements during sleep in weightlessness. Science. *225:* 221–222.

Reschke, M. F., D. J. Anderson, and J. L. Homick. 1984. Vestibulospinal reflexes as a function of microgravity. Science. *225:* 212–214.

Ross, H., E. Brodie, A. Benson. 1984. Mass discrimination during prolonged weightlessness. Science. *225:* 219–221.

Spangenburg, R. and D. Moser. 1987. A question of gravity. Space World. *X-2-278:* 8–11.

Variations in Human Physiology. (R. M. Case, Ed.). Manchester University Press. 1985.

Voss, E. W. 1984. Prolonged weightlessness and humoral immunity. Science. *225:* 214–215.

Young, L., C. M. Oman, D. G. D. Wattm, K. E. Money, B. K. Lichtenberg. 1984. Spatial orientation in weightlessness and readaptation to earth's gravity. Science. *225:* 205–208.

CHAPTER SEVEN

Brennan, P. J., G. Greenberg, W. E. Miall, S. G. Thompson. 1982. Seasonal variation in arterial blood pressure. Br. Med. J. *285:* 919–922.

Cohen, P. 1970. Seasonal variation of congenital malformations. J. Interdiscipl. Cycle. Res. *3:* 271–274.

Cowgill, U. M. 1966. Season of birth in man. Contemporary situation with special reference to europe and the southern hemisphere. Ecology. *47:* 614–623.

Falk, D. 1980. Hominid brain evolution: the approach from paleoneurology. Yrbook of Physical Anthropol. *23:* 93–107.

Fantel, A. G. 1978. Prenatal selection. Yrbook of Physical Anthropol. *21:* 215–222.

Findlay, A. L. R. *Reproduction and the Fetus.* University Park Press, Baltimore, Maryland. 1984.

Flexner, J. T. *Washington: the Indispensable Man.* Little, Brown and Company. Boston. 1974.

Hofman, M. A. 1984. On the presumed coevolution of brain size and longevity in hominids. J. Hum. Evol. *13:* 371–376.

Hopf, S. 1967. Notes on pregnancy, delivery, and infant survival in captive Squirrel Monkeys. Primates. *8:* 323–332.

Ingalls, A. J. and M. C. Salerno. *Maternal and Child Health Nursing.* C. V. Mosby Co., London. 1983.

Jongbloet, P. H. 1983. Menses and moon phases, ovulation and seasons, vitality and month of birth. Dev. Med. Child Neurol. *25:* 520–523.

Kalter, H. 1959. Seasonal variation in frequency of cortisone-induced cleft palate in mice. Genetics. *44:* 518–519.

Katz, G. 1953. Seasonal variation in incidence of premature births. Nord. Med. *50:* 1637–1638.

Kolata, G. 1984. Studying learning in the womb. Science. *225:* 302–303.

Leutenegger, W. 1974. Functional aspects of pelvic morphology in Simian primates. J. Hum. Evol. *3:* 207–222.

Lovejoy, C. O. 1981. The origin of man. Science. *211:* 341–350.

Lynch, G., S. Hechtel, and D. Jacobs. 1983. Neonatal size and evolution of brain size in the Anthropoid primates. J. Hum. Evol. *12:* 519–522.

Murray, A. B., A. C. Ferguson, and B. Morrison. 1980. The seasonal variation of allergic respiratory symptoms induced by house dust mites. Ann. Allergy. *45:* 347–350.

Pinatel, M. C., C. Souchier, J. P. Croze, J. C. Czyba. 1981. Seasonal variation of necrospermia in man. J. Interdiscipl. Cycle Res. *12:* 225–235.

Rosenblatt, L. S., M. Shifrine, N. W. Hetherington, T. Paglierone, M. R. MacKenzie. 1982. A circannual rhythm in rubella antibody titers. J.Interdiscipl. Cycle. Res. *13:* 81–88.

Scholten, C. M. *Childbearing in American Society: 1650–1850.* New York University Press, New York. 1985.

Shifrine, M., A. Garsd, L. A. Rosenblatt. 1982. Seasonal variation in immunity of humans. J. Interdiscipl. Cycle. Res. *13:* 157–165.

Smith, D. M., C. H. Conaway, and W. T. Kerber. 1978. Influences of season and age on maturation *in vitro* of rhesus monkey oocytes. J. Reprod. Fert. *54:* 91–95.

The Physiological Development of the Fetus and Newborn. (Jones, C. T. and P. W. Nathanielsz, Eds.). Academic Press, New York. 1985.

Wright, R. A., F. N. Judson. 1978. Relative and seasonal incidences of the sexually transmitted diseases—2 year statistical review. J. Vener. Dis. *54:* 433–440.

Ziegal, E. E. and M. S. Cranley. *Obstetric Nursing.* Macmillan Publishing Co., New York. 1984.

CHAPTER EIGHT

Beck, M. S. *Kidspeak.* Plume. New York. 1982.
Brain Development and Behavior. (M. B. Sterman, D. J. McGinty, and A. M. Adinolfi, Eds.) Academic Press. New York. 1971.
Brooks, V. and J. Hochberg. 1960. A psychophysical study of "cuteness". Percep. Motor Skills. *11:* 205.
Brown, J. L. *The Evolution of Behavior.* W. W. Norton and Co. New York. 1975.
Cowley, M. *The View from 80.* Penguin Books. New York, 1980.
Cutler, R. G. 1976. Evolution of longevity in primates. J. Human Evol. *5:* 169–202.
Darwin, C. *Expression of the Emotions in Man and Animals.* John Murray. London. 1872.
Dewhurst, J. *Female Puberty and its Abnormalities.* Churchill-Livingstone, New York. 1984.
Fries, J. F. 1980. Aging, natural death, and the compression of morbidity. New Engl. J. Med. *303:* 130–135.
Fullard, W. and A. M. Reiling. 1976. An investigation of Lorenz's "babyness". Child Devel. *47:* 1191–1193.
Gannon, L. R. *Menstrual Disorders and Menopause.* Praeger Scientific, New York. 1985.
Gardner, B. T. and L. Wallach. 1965. Shapes of figures identified as a baby's head. Percp. Motor Skills. *20:* 135–142.
Gassier, J. *A Guide to the Psycho-motor Development of the Child.* Churchill-Livingstone. New York. 1984.
Gedda, L. and G. Brenci. *Chronogenetics: The Inheritance of Biological Time.* Charles C. Thomas, Publ. Illinois. 1978.
Ginsburg, H. and S. Opper. *Piaget's Theory of Intellectual Development.* Prentice-Hall, Inc. New Jersey. 1969.
Gordon, I. *Forensic Medicine: a guide to principles.* Churchill-Livingstone, New York. 1983.
Gould, S. J. *The Panda's Thumb.* W. W. Norton and Co. New York. 1980.
Grumbach, M. M. 1980. The neuroendocrinology of puberty. In: *Neuroendocrinology.* (D. T. Krieger and J. C. Hughes, Eds.) Sinauer Assoc. Sunderland, Mass. pp. 249–258.
Hamilton, W. D. 1966. The moulding of senescence by natural selection. J. Theoret. Biol. *12:* 12–45.
Hayflick, L. 1970. Aging under glass. Exp. Geront. *5:* 291–303.
Kohn, R. R. *Principles of Mammalian Aging.* Prentice-Hall, New Jersey. 1978.
Kolata, G. 1984. Puberty mystery solved. Science. *223:* 272.
Lifting the Curse of Menstruation. (Golub, S., Ed.) The Haworth Press. New York. 1983.
Manning, A. *An Introduction to Animal Behavior.* Addison-Wesley Pub. Co. London. 1972.
Marshall, R. K. *Virgins and Viragos: A History of Women in Scotland from 1080 to 1980.* Academy Chicago Ltd. Illinois. 1983.
Pommerville, J. C. and G. D. Kochert. 1981. Changes in somatic cell structure during senescence of Volvox carteri. Europ. J. Cell Biol. *24:* 236–243.

Reitz, R. *Menopause: a Positive Approach.* Chilton Book Company. Radnor, Pennsylvania. 1977.
Review of Biological Research in Aging. Vol. 2. (M. Rothstein, Ed.), Alan R. Liss, Inc., New York. 1985.
Rothbart, M. K. 1973. Laughter in young children. Psychol. Bull. *80:* 247–256.
Stott, L. H. *Child Development: An Individual Longitudinal Approach.* Holt, Rinehart and Winston, Inc. New York. 1967.
Tanner, J. M. *Foetus into Man.* Harvard University Press, Massachusetts. 1978.
The Menopause. (H. J. Buchsbaum, Ed.) Springer-Verlag. New York. 1983.
Wickler, W. *The Sexual Code: the social behavior of animals and men.* Anchor Books. Garden City. 1978.

CHAPTER NINE

Ayala, F. J. 1987. Two frontiers of human biology: what the sequence won't tell us. Issues Sci. Tech. *3*(3): 51–56.
Barnes, D. H. 1987. Defect in Alzheimer's is on chromosome 21. Science. *235:* 846–847.
Bazell, R. Gene of the week. The New Republic. March 23, 1987.
Biotechnology & Biological Frontiers. (P. H. Abelson, Ed.) AAAS Publ. Washington, D.C. 1984.
D'Alessandro, A., J. H. Southard, M. Kalayoglu, and F. O. Belzer. 1986. Cryobiol. *23:* 161–167.
Davis, B. D. 1980. Frontiers of the biological sciences. Science. *209:* 78–89.
Dulbecco, R. 1986. A turning point in cancer research: sequencing the human genome. Science. *231:* 1055–1056.
Fahy, G. M. 1986. The relevance of cryoprotectant "toxicity" to cryobiology. Cryobiol. *23:* 1–13.
Fox, J. L. 1984. Injected virus probes fetal development. Science. *223:* 1377–1378.
Gilbert, W. 1987. Genome sequencing: creating a new biology for the twenty-first century. Issues Sci. Tech. *3*(3): 26–35.
Hood, L. and L. Smith. 1987. Genome sequencing: how to proceed. Issues Sci. Tech. *3*(3): 36–46.
Jacobsen, I. A. and D. E. Pegg. 1984. Cryopreservation of organs. Cryobiol. *21:* 377–384.
Kavaler, L. *Freezing Point.* John Day Co. New York. 1970.
Kneteman, N. M., R. V. Rajotte, and G. L. Warnock. 1986. Long-term normoglycemia in pancreatectomized dogs transplanted with frozen/thawed pancreatic islets. Cryobiol. *23:* 214–221.
Kolata, G. B. 1979. Developmental biology: where is it going? Science. *206:* 315–316.
Kolata, G. 1985. Genes and biological clocks. Science. *230:* 1151–2.
Kolata, G. 1982. Grafts correct brain damage. Science. *217:* 342–344.
Kolata, G. 1986. Researchers hunt for Alzheimer's disease gene. Science. *232:* 448–450.
Krauthammer, C. The ethics of human manufacture. The New Republic. May 4. 1987.

Lewin, R. 1986. Molecular biology of *Homo sapiens*. Science. *233:* 157–8.
Marx, J. L. 1987. Developmental control gene sequenced. Science. *236:* 26–27.
Marx, J. L. 1982. Transplants as guides to brain development. Science. *217:* 340–342.
Miller, J. A. 1985. Redesigning molecules nature's way. Science News. *128:* 204–206.
Pegg, D. E., I. A. Jacobsen, M. P. Diaper, and J. Foreman. 1986. Optimization of a vehicle solution for the introduction and removal of glycerol with rabbit kidneys. Cryobiol. *23:* 53–63.
Polge, C., A. U. Smith, and A. S. Parkes. 1949. Revival of spermatozoa after vitrification and dehydration at low temperatures. Nature. *164:* 666.
Schneider, K. Science debates using tools to redesign life. *The New York Times.* June 8, 1987.
Today a frozen dog, tomorrow the iceman. 1987. Discover. *8*(6): 9.
Watson, J. D. *Molecular Biology of the Gene.* W. A. Benjamin, Inc. London. 1976.
Zimmerman, M. R. 1985. Paleopathology in Alaskan mummies. American Scientist. *73:* 20–25.

APPENDIX

Cell Membranes: biochemistry, cell biology & pathology. (G. Weissmann and R. Claiborne, Eds), HP Publishing Co., Inc., New York. 1975.
Locke, D. M. *Enzymes—The agents of life.* Crown Publ., Inc. New York. 1969.
Molecules to Living Cells. W. H. Freeman and Co., San Francisco. 1980.
Rawn, J. D. *Biochemistry.* Harper and Row, Publ. New York. 1983.
Srivastava, D. K. and S. A. Bernhard. 1986. Metabolic transfer via enzyme-enzyme complexes. Science. *234:* 1081–1086.
Strand, F. L. *Physiology: a regulatory systems approach.* Macmillan Publ. Co. New York. 1983.
Stryer, L. *Biochemistry.* W. H. Freeman and Co. San Francisco. 1981.
Wolfe, S. L. *Biology of the Cell.* Wadsworth Publ. Co., Belmont, California. 1981.

INDEX